# AN INTRODUCTION TO ELECTRONICS

MAXWELL PRESS

# AN INTRODUCTION TO ELECTRONICS

**RALPH G. HUDSON**

Professor of Electrical
Engineering and Chairman of the Courses in General Science
and General Engineering at the Massachusetts
Institute of Technology

MAXWELL PRESS

Chennai Trichy NewDelhi

MAXWELL PRESS

This edition has been published in india by arrangement with Carson Books, UK

ISBN 978-93-90063-71-0 Maxwell Press

 No. 44, Nallathambi Street,
Triplicane, Chennai 600 005

MJP 915 

Publisher : C. Janarthanan

# Publisher's Note

The legacy of a country is in its varied cultural heritage, historical literature, developments in the field of economy and science. The top nations in the world are competing in the field of science, economy and literature. This vast legacy has to be conserved and documented so that it can be bestowed to the future generation. The knowledge of this legacy is slowly getting perished in the present generation due to lack of documentation.

Keeping this in mind, the concern with retrospective acquiring of rare books has been accented recently by the burgeoning reprint industry. Maxwell Press is gratified to retrieve the rare collections with a view to bring back those books that were landmarks in their time.

In this effort, a series of rare books would be republished under the banner, "Maxwell Press". The books in the reprint series have been carefully selected for their contemporary usefulness as well as their historical importance within the intellectual. We reconstruct the book with slight enhancements made for better presentation, without affecting the contents of the original edition.

Most of the works selected for republishing covers a huge range of subjects, from history to anthropology. We believe this reprint edition will be a service to the numerous researchers and practitioners active in this fascinating field. We allow readers to experience the wonder of peering into a scholarly work of the highest order and seminal significance.

**MAXWELL PRESS**

# AN INTRODUCTION TO ELECTRONICS

by Ralph G. Hudson

The enormous potentialities of the relatively new science of electronics have only begun to be used. It has given us radio, sound movies, facsimile reproduction, television. It has produced miracles in the use of radar, the radio compass, and other war equipment. Such instruments as the electron microscope are opening new vistas in the applied sciences, and there are still limitless possibilities in the application of electronics to the development of new materials and new sources of food and energy for man.

This book explains the science of electronics and its modern applications in terms that will be understandable and useful to those with only an elementary knowledge of mathematics and physics. It gives the reader a clear and scientifically exact knowledge of the modern theory of the constitution of matter and the nature of an electric current in a gas, a liquid, a solid, and in a vacuum. It describes and illustrates all of the major uses of electronic tubes and phototubes, the construction of electronic devices and their working principles.

*Broadcasting Company WNAX*

The tallest mast-type antenna in the United States is located at Yankton, South Dakota, and has an over-all height of 927 feet above the ground. Its total weight, including the down pull of six guy wires, is 190 tons which is supported by a single porcelain insulator, 30 inches in diameter and 36 inches high. For efficient radiation the height is 53 per cent of the wave length of the broadcasting frequency, which is 570 kilocycles per second.

# PREFACE

Electronics is that branch of science which describes the properties and control of electrons and other rudimentary particles which, in correlation with energy, constitute matter. Energy is the agency by which matter is united, disintegrated, or displaced, and appears in many diverse forms, such as mechanical, chemical, thermal, electrical, magnetic, and atomic energy. Matter is the substance of which any physical material is composed. The science might have been called "protonics" or "neutronics" with reference to other rudimentary particles but "electronics" was chosen to emphasize the prominence of the most active ingredient of matter.

Electronics has given man a better understanding of the hidden resources of nature. The discovery of the electron would still have attained enormous importance if no application of its properties had been made to radio communication, sound-motion pictures, or television. The science of electronics has opened a vista of applications which appear to be limitless and beyond our imagination. We have seen various applications which lend comfort, enlightenment, and amusement to mankind but these uses may pale into insignificance compared with the development of new materials and new sources of food and energy. The purpose in part of this book is to acquaint the layman with some of the outstanding new concepts of the physical world together with the implications of their probable effect upon his manner of living.

The understanding of any of the special or advanced sciences is greatly simplified by a preliminary study of physics, chemistry, and mathematics. Although such preparation gives the student a general knowledge of the basic principles of the fundamental sciences, a consequent attainment of greater importance is the acquirement of a scientific vocabulary. The author is always skeptical of the reason when a man says he "hasn't a scientific mind." It is more probable that he has not developed the habit of referring steadily to a comprehensive dictionary for the meaning of unfamiliar words. Beneficial reading of scientific literature requires not only an accurate understanding of scientific words but of common words as well.

Another handicap to accurate reading is the ascription of wrong meanings to important words. Thereafter the subsequent reading cannot make sense. Students of all grades often fail examinations not because they do not know the answers but because they do not know the exact meaning of the questions. The reason why many persons get nothing out of a legal decision, a doctor's report, or even an income-tax blank is because they do not know the meaning of the words. The first requirement for an adequate understanding of electronics without previous contact with the fundamental sciences is the possession and constant use of an up-to-date dictionary.

The lay reader will also experience some difficulty at first in comprehending some of the astronomical numbers that pervade the literature of electronics. For example, it is stated later in the text that "in the normal hydrogen atom the electron revolves about the proton nearly $10^{16}$ times in a second." The notation $10^{16}$ means one multiplied by 10, 16 times; that is, one with 16 zeros *after* it, or 10,000,000,000,000,000. When the reader becomes accustomed to this notation he will at once recognize $10^2$ as 100, $10^5$ as 100,000, $10^9$ as 1,000,000,000, and so on. Electronic literature also contains exceedingly small numbers. Referring to the proton the text states that "its diameter is about $10^{-16}$ centimeter," where $10^{-16}$ means one divided by one multiplied by 10, 16 times; that is, one with 16 − 1 zeros *before* it, to the right of the decimal point, or 0.000,000,000,000,000,1. Expressed as a fraction this number would be 1/10,000,000,000,000,000. In the simplified notation the reader will soon recognize $10^{-2}$ as 0.01, $10^{-5}$ as 0.000,01, and $10^{-9}$ as 0.000,000,001.

Aside from the above notation associated with large and small numbers the book contains practically no mathematics and assumes no previous study of physics or chemistry. In no instance have the difficult features of the evolution of electronics been neglected. Simple and complex features have all been included so that the reader will not feel that he has been let down. Most of the current popular literature on the subject tells what electronics "can" or "may do." In this exposition emphasis is placed not only on "can" or "may do" but on "how" and "why." Unless the reader has had considerable experience in reading scientific books he must not expect to read the book as he would a newspaper or a novel. He must read it slowly and perhaps several times. As the subject becomes clearer and clearer he will appreciate what Charles Kingsley, a minister, poet, and novelist, meant when he said "For science is . . . like virtue, its own exceeding great reward." Electronics will then become an intimate part of the reader's thought and experience.

It may also determine his life work. Vast industries have already been established to construct and operate electronic devices. A large part of the routine work in many research laboratories is dedicated to a study of electronic principles and their applications. Out of this research, industries now unknown may arise at any moment. Financiers must have sufficient knowledge of electronic fundamentals to decide whether each new project will succeed and deserve their support.

Many of these industries must have advertising personnel who will be able to write convincing copy. The doctor, the dentist, and the lawyer will discover that electronics to him should not be a strange science. Although it ranks first in the domain of the physicist, the chemist, and the electrical engineer, electronics has already become an essential science in various branches of civil engineering, mechanical engineering, metallurgy, biological engineering, chemical engineering, geophysics, marine engineering, and aeronautical engineering. Unquestionably electronics will not only establish its importance among most of the professions but also in the life of every man.

R. G. Hudson

*Cambridge, Massachusetts*

## TABLE OF CONTENTS

CHAPTER I

# THE CONSTITUTION OF MATTER

## The Cathode-ray Tube

When a voltage of sufficient magnitude from a battery or a generator is applied between the metal terminals of a partially evacuated glass tube, as shown in Fig. 1, a ray of light is established

**Fig. 1.** A luminous ray is produced within a glass tube containing a low-pressure gas when a high voltage is impressed between its terminals

between the negative terminal (the cathode) and the positive terminal (the anode). Although the luminous intensity of the ray is not high, this tube with certain modifications is nevertheless similar to the modern fluorescent lamp. A screen of any shape placed within the tube perpendicular to the ray produces a shadow of the same shape on the anode, thus indicating that something flowing within the tube from the cathode to the anode (and not the other way) is intercepted by the screen. The source of the ray being the cathode the tube is called a "cathode-ray" tube.

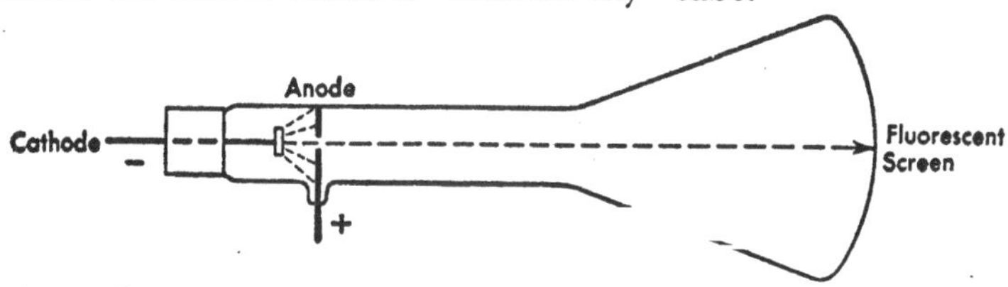


**Fig. 2.** Construction of the cathode-ray tube used with additions in a television receiver

In another type of construction, as shown in Fig. 2, the anode with a small hole drilled through its center is located a short distance from the cathode. With a voltage of at least 300 volts imthe inward pull of the proton. There were two serious objections to this theory. In the first place, there are an infinite number of radial distances and velocities that the electron might take which would provide the necessary compensation of forces. In the second place, it had heretofore been assumed that a revolving electron would steadily radiate energy and, in consequence, lose velocity and fall back into the proton.

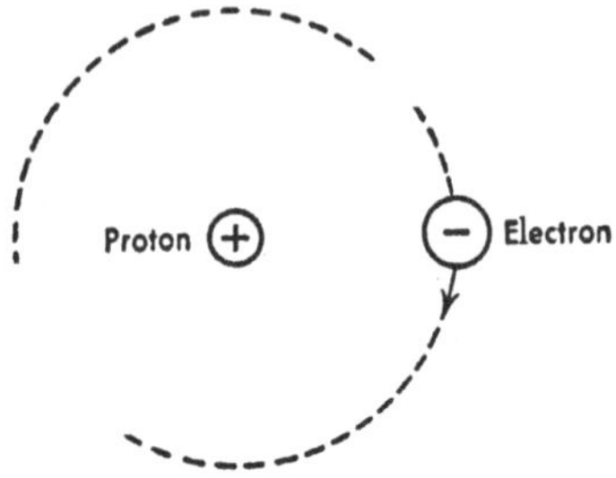


Fig. 3. A rudimentary model of the hydrogen atom

## The Bohr Model of the Atom

Evidence has accumulated in recent years that energy itself is not divisible into any number of parts, but that the division is limited to certain portions, under the prescribed circumstances, called "quanta." A quantum of energy is not a fixed quantity like the charge on an electron or a proton but varies with the conditions to which it must react. Applying this principle to the hydrogen atom, Dr. Niels Bohr suggested that the electron may revolve about the proton, as shown in Fig. 3, at definite radial distances and velocities but that it takes a certain indivisible quantum of energy to make it spiral from one orbit to another. If an insufficient amount of energy is received or lost it must remain where it is.

According to this theory, in the normal hydrogen atom the electron revolves about the proton nearly $10^{16}$ times in a second at a radial distance of about $10^{-8}$ centimeter and with a velocity of about 4000 miles per second. The system of units is scrambled because it is believed that the reader may visualize miles per second better than centimeters, meters, or kilometers per second. A model of the hydrogen atom, or any other atom, may not be drawn to scale. If the electron is made sufficiently large to be seen, the proton would be invisible and off the page. Other theories of atomic structure have been proposed at various times but at the moment the Bohr model has proved quite satisfactory in presenting the simplest aspect of the picture. Greater refinement of atomic structure indicates that an electron in motion must be regarded not only as a particle but also as a wave.

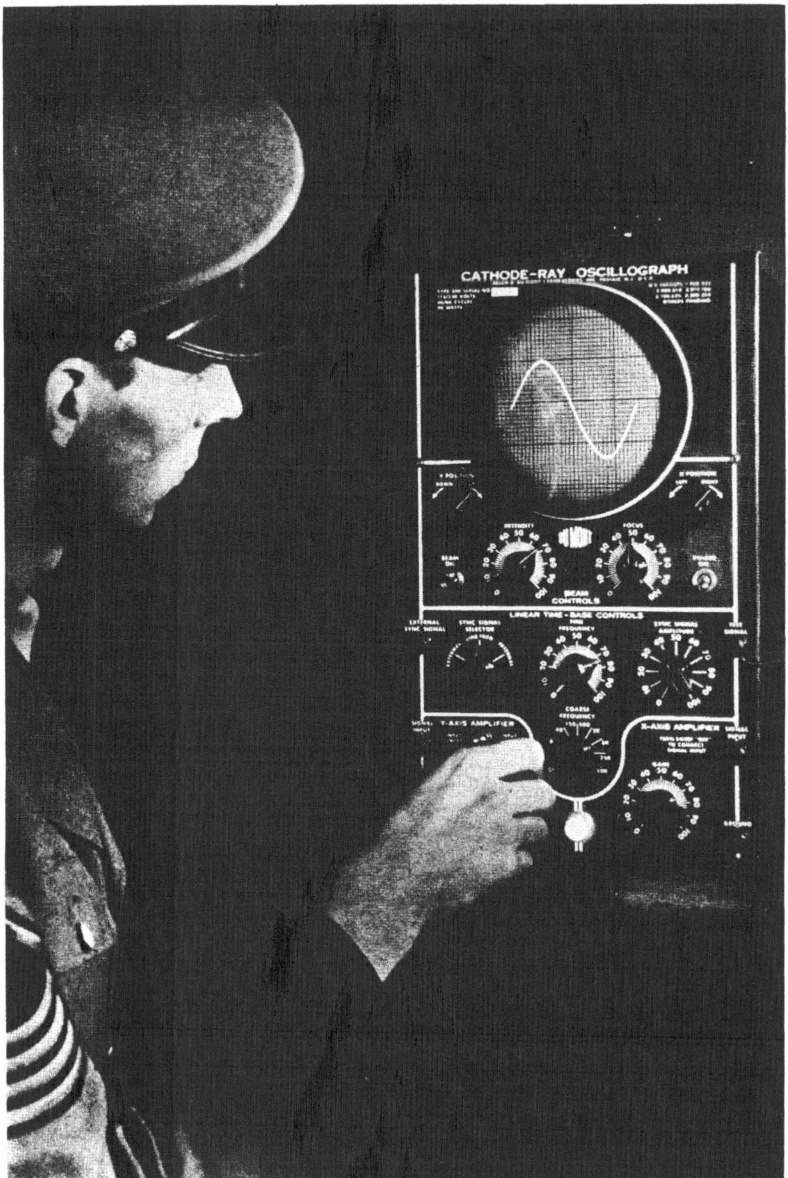


*Allen B. Dumont Laboratories, Inc.*

A cathode-ray oscillograph which traces on a fluorescent screen the instantaneous variations of an electric current. The picture shows a sinusoidal alternating current.

*Dean G. R. Harrison and associates at the Massachusetts Institute of Technology*

View inside a large grating-spectrograph at M.I.T. Light from the activated material, which enters through a slit located in the opposite wall, is broken up into its constituent wave lengths by a diffraction grating (not visible) at the extreme right. The resulting spectrum lines are photographed on plates arranged around the circular track which is 21 feet in diameter. The room is kept to within 0.01 degree Fahrenheit of a definite temperature, usually 65 degrees.

## The Helium Atom

The next heavier atom among the elements is helium which weighs about four times as much as the hydrogen atom. To give the nucleus the proper mass it was first assumed that it contains four protons and two electrons, surrounded by two revolving electrons. Subsequent bombardment of helium and various other gases with high-velocity particles has shown that when a nucleus

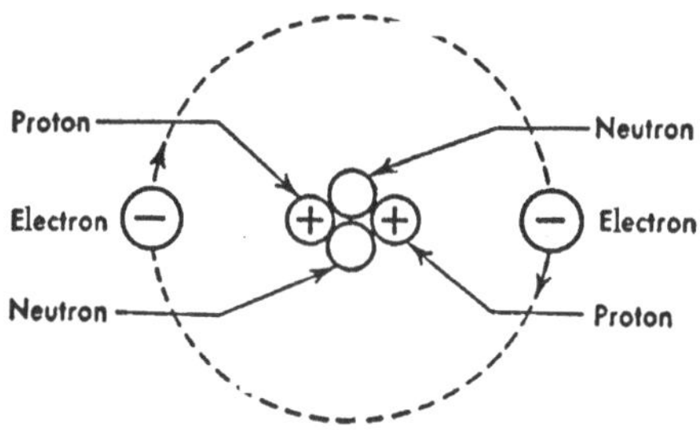


**Fig. 4.** The constituents of the helium atom

is disrupted it often throws out a particle which has no charge and has a mass slightly greater than that of a proton. This particle has appeared in many nuclear reactions and is called a "neutron." It is now definitely established that the helium nucleus contains two protons and two neutrons, which combined account for almost the entire weight of the atom. This model of the helium atom (not to scale) is shown in Fig. 4. Since all other heavier atoms have also shown evidence of possessing neutrons in their nuclei, it is believed that the three stable building blocks of the universe are the electron, the proton, and the neutron.

## The Spectroscope

It is not possible in this brief and nonmathematical discussion to explain how Bohr and others have determined the nuclear and orbital structure of the 92 elements from hydrogen with an atomic weight of about 1 to uranium with an atomic weight of about 238. Much of the contributory evidence has come from the spectroscope. In this instrument a small piece of an element to be examined is dropped into the high-temperature crater of a carbon-arc lamp where the individual atoms of the element are activated and throw out visible and invisible radiation of frequencies characteristic of the particular atom. This radiation is sent either through an optical prism or against an opaque grating of many thousands of equally spaced lines. In either case a spectrogram obtained photographically or visually (for part of the spectrum) will show the various frequencies of the emitted radiant energy.

## The Electron Orbits

The energy stored in the revolving electron is the least when it is located in the innermost orbit. If activation of some sort, such as the intense heat of a carbon arc, drives the electron into an outer orbit, it will spiral back to an inner orbit and give out high-frequency radiation during the passage. An electron moves from an inner to an outer orbit only upon the absorption of energy from some external source. This energy is radiated wholly or partly almost immediately when the electron falls back to an inner orbit, and upon arrival at an inner orbit all radiation ceases. As shown in a later chapter practically all sources of light operate on this principle. The amount of energy radiated when an electron spirals inward equals $6.55 \times 10^{-27}$ times the frequency of the radiant energy. Since the spectrograph indicates the energy differences between characteristic orbits for various atoms it is possible with other information to determine the number and location of the orbits of any atom

## The Bohr Model in General

Many attempts have been made to make models or draw pictures showing the distribution of the orbital electrons in the various atoms. In an early model (Fig. 5 *a*) the atomic structure is represented by an arrangement of the electrons in seven concentric shells, like the layers of an onion, in which each shell represents a quantum level one unit less than the next outer shell. The shells are not equally spaced but progress outward with steadily increasing spaces; the shells moreover are not spheres but ellipsoids. This model was soon superseded by a more detailed and diaphanous structure (Fig. 5 *b*) containing various elliptic orbits arranged in a symmetrical pattern about the nucleus. Each orbit, while not necessarily concentric to the others, is located in one of several quantum levels.

This model again proved to be too simple and restrictive to satisfy the observed facts and at the moment we can draw no definite picture of the distribution of the electrons about the nucleus. The failure to establish a satisfactory picture does not indicate a reversal of progress or that we know less than before about the structure of the atom. On the contrary it means that we know more. The

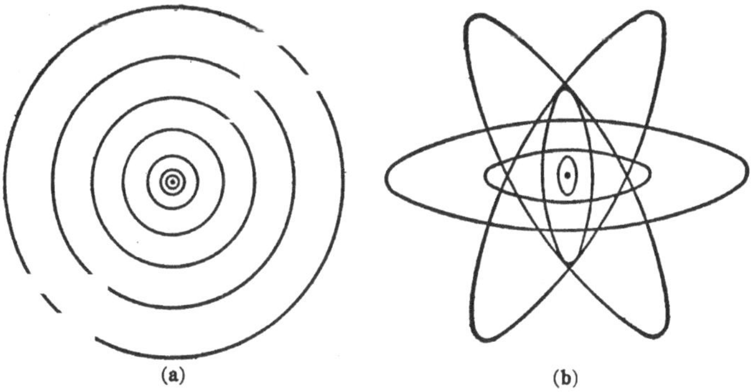


**Fig. 5. a.** The concentric-shell model of the atom. **b.** The symmetrical elliptical-orbit model of the atom

model is now described by a series of mathematical equations which are more accurate and descriptive than a picture. Modern science in several instances has developed explanations of physical phenomena which do not suggest pictures. As yet nobody has been able to make a model containing a fourth dimension, but the failure of this accomplishment does not disprove its existence. Improvement in the understanding of the structure of atoms will continue without the necessity of a picture. We may even use some of the old ones at times because they do contain some of the characteristics of the real structure.

## The Ninety-two Elements

The table on page 9 gives the name, the symbol, the atomic weight, and the atomic number or the number of orbital electrons for the atoms of the 92 elements. Where there are blank spaces the properties of that element are either unknown, or the element itself has not been discovered or isolated. There are also indications that two other elements may exist which are heavier than uranium, thus making it possible that there are actually 94 elements. The atomic weights are comparative values based upon oxygen as the standard with an atomic weight of exactly 16. If hydrogen had been taken as the standard the atomic weight of oxygen would be 15.871.

Frequent references will be made later regarding the meaning and interpretation of this table so that it will not appear to be just a jumble of numbers. Every living thing, so far as we know, contains carbon, and that atom contains six orbital electrons. Hence that arrangement is a requisite of life. It will be noted that iron, cobalt, and nickel have 26, 27, and 28 orbital electrons respectively. These arrangements appear to be requisites for highly magnetic materials. All the elements from bismuth through uranium are radioactive; that is, they disintegrate steadily and change into elements with fewer orbital electrons. Let us now make a study of the atomic weights.

### Atomic Numbers and Isotopes

When we have spoken of the probable existence of 92 elements (perhaps 94) we have meant uncombined substances which possess unique properties and occur in nature with four possible exceptions, in sufficient quantity to be detected. The better understanding of the constitution of matter has enabled us to discover hundreds of other different atoms whose existence might otherwise have remained unknown. Since all atoms are made of an equal and increasing number of protons and electrons combined with neutrons, it was difficult to explain why the table of atomic weights on page 9 contains many gaps in their orderly sequence. The first four elements, for example, have atomic weights as follows: hydrogen, 1.0080; helium, 4.003; lithium, 6.940; and beryllium, 9.02. In a simple atomic evolution why were there no elements with atomic weights of 2, 3, 5, 6, and 8? Also why were the atomic weights not integers? According to the Bohr model a hydrogen atom contains one proton and one electron and the oxygen atom contains eight protons, eight neutrons (each neutron weighing about the same as one proton), and eight electrons. Yet the oxygen atom weighs less than 16 times that of the hydrogen atom.

An examination of the Bohr models for all the elements shows that the number of orbital electrons, called the "atomic number," in any atom equals roughly (in some cases closely) one half of the atomic weight. Aston suggested that the wide departure from this rule in the case of lithium, for example, with an atomic weight of

# THE NINETY-TWO ELEMENTS

| *Name* | *Symbol* | *Atomic Weight* | *Atomic Number or Number of Orbital Electrons* | *Name* | *Symbol* | *Atomic Weight* | *Atomic Number or Number of Orbital Electrons* |
|---|---|---|---|---|---|---|---|
| Hydrogen | H | 1.0080 | 1 | Silver | Ag | 107.880 | 47 |
| Helium | He | 4.003 | 2 | Cadmium | Cd | 112.41 | 48 |
| Lithium | Li | 6.940 | 3 | Indium | In | 114.76 | 49 |
| Beryllium | Be | 9.02 | 4 | Tin | Sn | 118.70 | 50 |
| Boron | B | 10.82 | 5 | Antimony | Sb | 121.76 | 51 |
| Carbon | C | 12.010 | 6 | Tellurium | Te | 127.61 | 52 |
| Nitrogen | N | 14.008 | 7 | Iodine | I | 126.92 | 53 |
| Oxygen | O | 16.0000 | 8 | Xenon | Xe | 131.3 | 54 |
| Fluorine | F | 19.00 | 9 | Cesium | Cs | 132.91 | 55 |
| Neon | Ne | 20.183 | 10 | Barium | Ba | 137.36 | 56 |
| Sodium | Na | 22.997 | 11 | Lanthanum | La | 138.92 | 57 |
| Magnesium | Mg | 24.32 | 12 | Cerium | Ce | 140.13 | 58 |
| Aluminum | Al | 26.97 | 13 | Praseodymium | Pr | 140.92 | 59 |
| Silicon | Si | 28.06 | 14 | Neodymium | Nd | 144.27 | 60 |
| Phosphorus | P | 30.98 | 15 | —— | — | —— | 61 |
| Sulphur | S | 32.06 | 16 | Samarium | Sm | 150.43 | 62 |
| Chlorine | Cl | 35.457 | 17 | Europium | Eu | 152.0 | 63 |
| Argon | A | 39.944 | 18 | Gadolinium | Gd | 156.9 | 64 |
| Potassium | K | 39.096 | 19 | Terbium | Tb | 159.2 | 65 |
| Calcium | Ca | 40.08 | 20 | Dysprosium | Dy | 162.46 | 66 |
| Scandium | Sc | 45.10 | 21 | Holmium | Ho | 163.5 | 67 |
| Titanium | Ti | 47.90 | 22 | Erbium | Er | 167.2 | 68 |
| Vanadium | V | 50.95 | 23 | Thulium | Tm | 169.4 | 69 |
| Chromium | Cr | 52.01 | 24 | Ytterbium | Yb | 173.04 | 70 |
| Manganese | Mn | 54.93 | 25 | Lutecium | Lu | 174.99 | 71 |
| Iron | Fe | 55.85 | 26 | Hafnium | Hf | 178.6 | 72 |
| Cobalt | Co | 58.94 | 27 | Tantalum | Ta | 180.88 | 73 |
| Nickel | Ni | 58.69 | 28 | Tungsten | W | 183.92 | 74 |
| Copper | Cu | 63.57 | 29 | Rhenium | Re | 186.31 | 75 |
| Zinc | Zn | 65.38 | 30 | Osmium | Os | 190.2 | 76 |
| Gallium | Ga | 69.72 | 31 | Iridium | Ir | 193.1 | 77 |
| Germanium | Ge | 72.60 | 32 | Platinum | Pt | 195.23 | 78 |
| Arsenic | As | 74.91 | 33 | Gold | Au | 197.2 | 79 |
| Selenium | Se | 78.96 | 34 | Mercury | Hg | 200.61 | 80 |
| Bromine | Br | 79.916 | 35 | Thallium | Tl | 204.39 | 81 |
| Krypton | Kr | 83.7 | 36 | Lead | Pb | 207.21 | 82 |
| Rubidium | Rb | 85.48 | 37 | Bismuth | Bi | 209.00 | 83 |
| Strontium | Sr | 87.63 | 38 | Polonium | Po | 210 | 84 |
| Yttrium | Y | 88.92 | 39 | —— | — | —— | 85 |
| Zirconium | Zr | 91.22 | 40 | Radon | Rn | 222 | 86 |
| Columbium | Cb | 92.91 | 41 | —— | — | —— | 87 |
| Molybdenum | Mo | 95.95 | 42 | Radium | Ra | 226.05 | 88 |
| —— | — | —— | 43 | Actinium | Ac | 227 | 89 |
| Ruthenium | Ru | 101.7 | 44 | Thorium | Th | 232.12 | 90 |
| Rhodium | Rh | 102.91 | 45 | Protactinium | Pa | 231 | 91 |
| Palladium | Pd | 106.7 | 46 | Uranium | U | 238.07 | 92 |

6.940 and an atomic number of 3, might be due to a mixture of two kinds of lithium with atomic weights of 6 and 7, respectively. A careful search for these two types of lithium showed that they actually exist and they are called the "isotopes" of lithium. The first isotope has a nucleus containing three protons and three neutrons surrounded by three orbital electrons, while the second isotope has a nucleus containing three protons and four neutrons surrounded by three orbital electrons. Since the chemical and physical properties of an atom, aside from its atomic weight and radioactivity, depend entirely upon the number of orbital electrons, each of these isotopes possesses essentially the same properties. In nature they normally appear in an aggregate of (let us say) 100 atoms of lithium with 6 atoms of the first isotope and 94 of the second, making the combined atomic weight, 6.940.

Investigation of the other gaps led to the discovery that all the elements possess two or more isotopes, some of which are stable and others radioactive; that is, changing into some other form. In the case of hydrogen it was found that in addition to the usual form of hydrogen with an atomic weight of about 1, there are two other isotopes with atomic weights of about 2 and 3, respectively. The first isotope, called "protium," has a nucleus of one proton centered on one orbital electron; the second isotope, called "deuterium," has a nucleus of one proton and one neutron centered on one orbital electron; the third isotope, called "triterium," has a nucleus of one proton and two neutrons centered on one orbital electron. In a molecule of ordinary water containing two atoms of protium and one atom of oxygen, the molecular weight is $2 \times 1 + 16$ or 18. If deuterium is substituted for the protium the molecular weight is $2 \times 2 + 16$, or 20. The latter liquid is called "heavy water." It differs in many ways from ordinary water. For example, it freezes at 3.08 degrees Centigrade instead of zero and boils at 101.4 degrees instead of 100 degrees. Small fish are killed in a few hours when transferred from ordinary water to heavy water.

While the number of different atoms was thought a few years ago to be 92, the addition of the different isotopes has increased the number in a recent count to 600. It will be seen that nature does not in fact leave gaps in the orderly sequence of the atomic weights. It appears probable that the few gaps that still remain will eventually be filled. As the determination of atomic weights by different methods becomes more accurate it has become certain that their values will not be integers and for a significant reason, although the departure from integral values is very small. Although all the heavier atoms contain a certain multiple of the constituents of the hydrogen atom, the heavier atom cannot weigh that multiple times the weight of the hydrogen atom because the formation of any nucleus is accompanied by a conversion of some of the mass into energy.

## Relation between Mass and Energy

It was predicted in Einstein's theory of relativity that mass and energy are interchangeable; that is, mass is a form of energy, and vice versa. In the formation of the nucleus of one of the heavy atoms a great many protons and neutrons are packed into a space not much larger than that occupied by a single electron. The energy required to accomplish this packing comes from a loss in mass of the constituents. According to Einstein this loss of mass in grams equals the energy radiated in ergs divided by the square of the velocity of light in centimeters per second. In consequence of this rule, when two or more electrically charged particles are packed into a very small space the resultant mass must be slightly less than the sum of the constituent masses. The deuteron, for example, weighs slightly less than the sum of the weights of a proton and a neutron.

## The Wilson Cloud Chamber

Evidence has been presented that all substances are built of electrons, protons, and neutrons. There is further evidence that other unique particles do exist but only for an extremely short time. One of the simplest experimental devices used in the investigation of rudimentary particles is the Wilson cloud chamber. It consists of a flat cylindrical box, with a tight glass cover, containing water or alcohol vapor, as shown in Fig. 6 *a*. A piston pump or a rubber diaphragm is connected to the inside of the chamber so that the pressure of the vapor may be suddenly reduced. Under ordinary conditions the water vapor is invisible, as it is in a cloudless sky, but when the pressure is reduced (a falling barometer foretells rain) droplets of water condense on any charged particles within the chamber. A moving charged particle will leave a trail of droplets behind it and its path may be seen or photographed. The box is furthermore placed between the poles of a magnet so that magnetic flux is sent through the box as shown in the figure. If a proton passes through the chamber its path (viewed from above and to the right) will curve to the right, as shown in Fig. 6 *b*, and an electron will curve to the left. A neutron, being uncharged, will

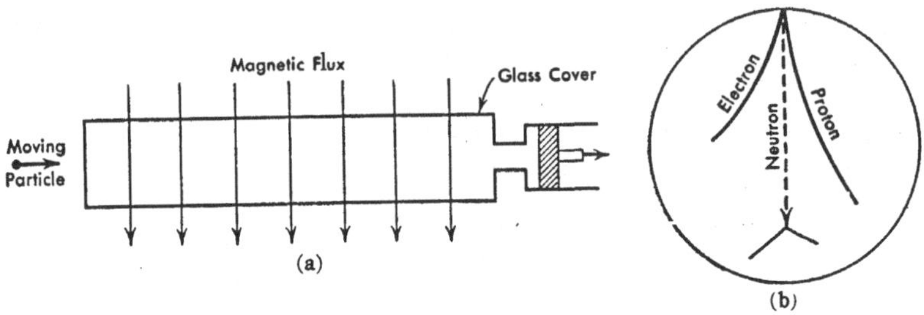


**Fig. 6. a.** The Wilson cloud chamber. **b.** Rudimentary particles identify themselves in the cloud chamber

leave no path but will often manifest itself by striking some other particle and splitting it into two charged particles, one or both of which may be accelerated and leave a path as shown in the figure. The Wilson cloud chamber, aside from indicating whether a partiele passing through it is positively or negatively charged, will also indicate, by the length and curvature of the path and the density of the track, the mass, charge, and velocity of the particle.

## Cosmic Rays, Positrons, and Mesotrons

It has been known for some time that the earth is bombarded steadily by powerful penetrating rays, called "cosmic rays." It is not known definitely what they are but it is generally believed that they contain in part high-velocity particles. When a cosmic ray strikes any object near a Wilson cloud chamber it sends out by impact many secondary charged particles, the paths of which may be photographed in the chamber. In one of the photographs which has been duplicated many times it was found that the right-hand track in Fig. 6 *b*, although curved to the right like the proton track, was otherwise the exact duplicate of the electron track on the left. This indicated that the cosmic ray had produced a positive particle with the same mass and charge as the electron. This new particle, called a "positron," is only produced when an object is bombarded by a cosmic ray or other high-velocity particles. It appears on the scene only for a small fraction of a second and then combines with any free electron. The combination results in the destruction of both masses which are converted into an equivalent amount of energy. In a similar manner the Wilson cloud chamber has shown the track of a particle, sometimes positive and sometimes negative, which has a mass about 200 times that of the electron. This particle, called a "mesotron," changes in a few millionths of a second into products not definitely known.

### Balance of Energy Requires a Neutrino

During the disintegration of certain radioactive materials the amount of energy ejected is often less than the amount of energy lost by the material. It is believed that the discrepancy may be due to the undetected ejection of an uncharged particle of very small mass, called a "neutrino." This particle has never been isolated but has a name ready for it the instant it appears. It may be said that the addition of positrons, mesotrons, and neutrinos to the stable particles already described does not seriously complicate the electronic world because they seldom enter the picture and then only as a flash in the pan.

### Disintegration of the Elements

It is common knowledge that elements with atomic weights exceeding that of lead are unstable and disintegrate into elements of lower atomic weights which are stable such as lead, itself. More recently it has been observed that some of the isotopes of lighter elements, such as potassium, rubidium, samarium, and lutecium, also undergo reductions in atomic weight by the same process. Disintegration in each case is accompanied by the emission of particles and rays, all originally called "$\alpha$ (alpha)," "$\beta$ (beta)," and "$\gamma$ (gamma)" rays. It is now definitely known that the $\alpha$-ray consists of the high-velocity nuclei of helium atoms, as shown in Fig. 4. Their velocity ranges from 8000 to 12,000 miles per second. These nuclei are also called "alpha particles" and contain two protons and two neutrons. The $\beta$-ray consists of high-velocity electrons or positrons having a velocity approaching that of light, 186,000 miles per second. Since both $\alpha$-rays and $\beta$-rays consist of moving charged particles, they are easily identified in many ways; for example, in a Wilson cloud chamber.

## The Photon

The $\gamma$-ray consists of very high-frequency pulses of radiant energy, each pulse or quantum of energy being called a "photon." The $\gamma$-ray produces only indirect effects in the Wilson cloud chamber, such as the splitting of obstructing particles, but manifests itself directly by the photoelectric emission of electrons from certain surfaces upon which it may impinge. This phenomenon of photoelectric emission is discussed in detail in another chapter.

## Electronics Comes to the Aid of Alchemy

The fact that radioactive elements are constantly changing from one element to another in nature has suggested the possibility of effecting such transmutation artificially and thereby realizing the dream of the unsuccessful alchemists of old. Bombardment of the elements, usually in gaseous form, by any of the rays will easily remove one or more of the orbital electrons from an atom, but that type of disintegration is only transitory because other free electrons are picked up immediately and the atom becomes normal again. Actual transmutation is obtained only by removing or adding something to the nucleus itself.

Since the formation of a nucleus is accomplished only by the expenditure of a large amount of energy, its disruption will require a proportionately large amount. No way has been found to gain contact with an individual nucleus and change its structure other than to bombard an element in the aggregate with extremely high-velocity particles of large mass, such as neutrons, $\alpha$-particles, or deuterons, the nuclei of deuterium. Some of the most ponderous machines of the modern physical laboratory have been constructed to produce these ultrahigh velocities. Since the nucleus of any atom probably does not exceed $10^{-12}$ centimeter in diameter it will be realized that very few of the bombarding particles will make a direct hit on a nucleus. Any large-scale transmutation of the elements by the rays of natural radioactive materials is easily shown to be futile. The strongest radioactive materials, like radium, in an amount not too expensive to obtain, send forth about $10^9$ $\alpha$-particles per second, and not more than one nuclear hit may be expected for each $10^6$ $\alpha$-particles emitted. Under the circumstances at the beginning of the bombardment about $10^3$ atoms may be transmuted per second. Since a gram of matter of any kind contains about $10^{22}$ atoms the transmutation of the entire amount would take at least $10^{22}$ divided by $10^3$ or $10^{19}$ seconds. This is about one billion years.

## The Electron-Volt

A convenient unit of measurement of the disrupting energy associated with a high-velocity particle is the "electron-volt." Suppose we should hold an electron with some insulated tongs just under the positive wire of a direct-current, 115-volt house circuit. The positive wire would attract the electron and the negative wire would repel it. It would therefore require work, or the expenditure of energy to force the electron from one wire to the other. This energy would be stored in the electron (and the surrounding electrostatic field) and its magnitude would be called "115 electron-volts." If we should release it at the negative wire it would fly back to the positive wire. If it were stopped completely near the positive wire by impact with the nucleus of an atom it would give up to that nucleus 115 electron-volts of energy. Since one electron-volt is the equivalent of only $1.6 \times 10^{-12}$ erg it will be seen that a particle must be given an energy of millions of electron-volts to do much damage by collision with another particle. Mega-electron-volts or a million electron-volts is abbreviated "mev." Most of the machines now in use develop several mev in their accelerated particles and it is expected that some machines now under construction will produce high-velocity particles with a disruptive energy of 100 mev.

## The Cyclotron

In the Lawrence cyclotron, two semicircular hollow metal boxes, called "dees," about 60 inches in diameter, are separated by a gap, as shown in Fig. 7, and are placed within and insulated from a tight circular box so that the dees may be immersed in any gas at any degree of vacuum.

An incandescent filament heated by a high-frequency current and located at the center of the dees will break up or ionize the atoms of the surrounding gas into two oppositely charged particles; in the case of deuterium, for example, into the positively charged deuteron (one proton and one neutron) and one negatively charged electron. An oscillator (described in another chapter) with a voltage of about 10,000 volts and a frequency of $10^7$ cycles per second is connected between the two dees. When the right-hand dee is negative a deuteron, which is positive, will be pulled and accelerated toward the right-hand dee. Since the dees and the enclosing box are located between the poles of a very powerful magnet with the magnetic flux perpendicular to the flat surfaces of the dees, the

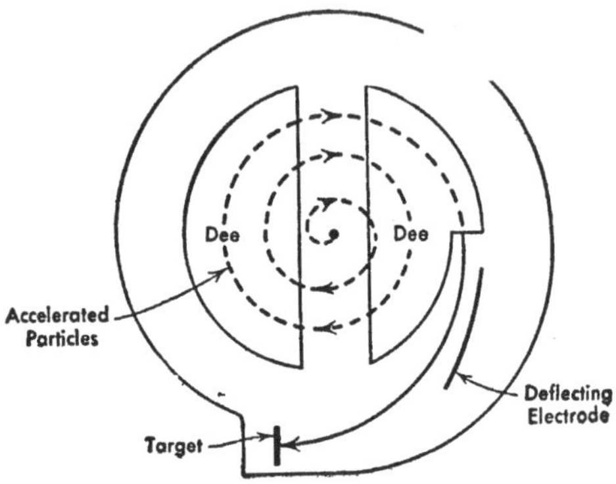


**Fig. 7.** The cyclotron

accelerated deuteron will describe a half-circle within the right-hand dee. When the deuteron arrives at the inner boundary of the right-hand dee the oscillator reverses the voltage between the dees, making the left-hand dee negative, and accelerates the deuteron toward the left-hand dee where it describes a larger half-circle at increased velocity. The deuterons thus spiral outward until they approach a high-voltage, negatively charged deflecting plate and then pass through an opening at an exceedingly high velocity. The nuclei of the atoms of any gas placed near the opening will be disrupted by this bombardment so that the original atoms will be transmuted into others, usually by the loss of neutrons which escape into the surrounding space. Cyclotrons are usually surrounded by tanks of water about 4 feet thick to stop radiated neutrons which would be injurious to the operators, who also never come nearer than 40 feet from the cyclotron while under operation. The new cyclotron being built at the University of California will have a pole-face diameter of 184 inches. The magnet will be 30 feet high and 56 feet long, and will weigh 4000 tons.

## Other Sources of High-energy Particles

In the Kerst induction accelerator, called a "betatron," an evacuated doughnut-shaped glass chamber, 5 feet in diameter, forms the single-turn secondary of a 600-cycles per second alternating-current transformer. When the alternating magnetic flux linking the accelerating chamber begins to increase from zero in one of its cycles, electrons are injected through a port at one point on the outer surface of the chamber. The electrons are then accelerated by the induction principle applying to any transformer secondary but are not allowed to alternate. When the linkage flux in the particular cycle reaches its maximum value the distribution of the flux density in the chamber is changed suddenly so that the high-velocity electrons are thrown through an opening in the periphery of the ring. In the quarter-cycle (1/2400 second) impulse, the electrons travel around the chamber 200,000 times and reach a velocity of 18,600 miles per second or one tenth the velocity of light. Each electron then possesses an energy of 20 mev. They are usually made to impinge upon the heavy metal target of a highly evacuated X-ray tube. The ray emitted by the target in consequence of this bombardment possesses exceedingly high penetrating energy and is used principally for photographing the interior of objects opaque to ordinary light. In accelerating electrons, the betatron supplies a function not available in the cyclotron, because electrons accelerated in the cyclotron change mass rapidly at extremely high velocities and fall out of step with the voltage alternations between the dees.

In the Van de Graaff electrostatic generator a hollow aluminum sphere, 15 feet in diameter, is located at the top of a hollow column of insulating material, 22 feet in height and 6 feet in diameter. A continuous rubber belt, 47 inches wide, runs over two pulleys, one near the ground and the other just inside the sphere. The lower pulley is driven by an electric motor so that the belt runs at a velocity of 5650 feet per minute. At the bottom of the upgoing side of the belt positive charges from a 100,000-volt source of direct current are sprayed on the belt from a long metal comb. These charges are taken off the belt at the top by a similar metal comb, or collector, and transferred to the sphere. It will be seen that positive charges must steadily accumulate on the sphere and that the voltage is limited only by the amount which will eventually discharge through the air to the ground. By operating the generator in a special building under pressure of an insulating gas (Freon) a voltage of 2.5 million volts has been obtained. This voltage applied to any positively charged particles placed at the top of a long evacuated tube causes them to accelerate rapidly down the tube and produce high nuclear disruption in any substance which they may strike at the bottom.

### Transmutation Achieved

Most of the elements have been transmuted by bombardment into elements of lower or higher atomic weight or into one of their isotopes. In most cases the amount of material transmuted is very small and is detected only by evidence of its radioactivity. If we wish to make gold (see page 9), we may bombard mercury (atomic weight, 200.61) and expel from the nucleus of each atom one proton and two neutrons so that the atomic weight will be reduced to that of gold, 197.2, and with an atomic number of 79. This has been accomplished on a very small scale but the resultant gold was an unstable isotope. Another process would involve the addition of one proton and one neutron to the nucleus of a platinum atom (atomic weight, 195.23), and we should again have an atom of gold. Since the original platinum is worth more than the gold this would not be a profitable business. Although it is cheaper at the moment to dig for gold than to transmute it in the cyclotron, we nevertheless know that more powerful sources of high-velocity particles may be constructed which may enable us to transmute common elements into rare elements in large quantities. The cyclotron already has produced considerable amounts of mildly radioactive materials which have been used in place of the more expensive radium to eliminate malignant growths. These materials are not only cheaper but, having a short radioactive life, may be inserted surgically in the malignant tissue and allowed to disappear with the malignant growth. When radium is used for the same purpose it must be inserted in one operation and removed in another.

### Utilization of Atomic Energy

In the preceding pages frequent mention has been made of the energy radiated by loss of mass when two or more particles combine to form a nucleus. There has been much speculation as to the possibility of utilizing this energy for practical purposes. This plan is particularly alluring because the amount of atomic energy to be obtained from very small masses of matter is stupendous. The carbon and hydrogen contained in a pound of bituminous coal, represented by a cube about 2.7 inches on a side, when combined molecularly in perfect combustion with oxygen would release 4.4 kilowatt-hours of energy. A kilowatt-hour may be visualized as the work done in carrying 133 tons of material up a stairway rising 10 feet and is the unit of energy upon which our electric-light bills are based. If, in comparison, we assume this pound of coal to be converted atomically, the Einstein equation shows that the energy released would be over 10 billion kilowatt-hours and, at 5 cents per kilowatt-hour, would be worth $500,000,000. It may be said at once that we have no conception of a method for changing all the mass of any substance into energy.

If we review the fact already described, that the formation of a nucleus by combining protons and neutrons is accompanied by a slight loss of mass and the release of a small amount of energy, that plan might appear to have some merit. To obtain the protons we could buy a flask of hydrogen, but where could we obtain a flask of neutrons? We can, to be sure, obtain transitory neutrons by bombardment, but we not only would obtain too few for the effort but we do not know how to store them in a flask because they combine with adjacent protons as fast as they are produced. Another plan involves the bombardment of a mixture of lithium and hydrogen, for example, with protons and with the hope that the emitted $\alpha$-particles will in turn disrupt other adjacent lithium atoms and release a steadily increasing amount of energy just as we kindle a firepot of coal with paper and wood. This project also does not work because the nucleus of an atom is well protected against intrusion by the surrounding shells of electrons. When it is realized that the volume occupied by the orbital electrons of any atom is about a million million times as great as the volume of the nucleus, it will be seen that practically all the $\alpha$-particles emitted by the lithium will be stopped in one of the outer shells of an adjacent atom before they reach its nucleus.

Finally, it has been observed that the bombardment of uranium, especially some of its isotopes, with neutrons will split the uranium atom into two atoms of elements of about the same lighter atomic weight. This form of disruption is called "fission." In this peculiar process, which applies as well to other heavy elements, the atoms are practically cut in two with the emission of a large amount of energy resulting from a loss in mass. The effectiveness of the bombardment with neutrons is due to the fact that they are uncharged and pass through the outer shells with little opposition. It has been computed that 1 cubic meter of uranium oxide might be expected to release $10^{12}$ kilowatt-hours in less than 0.01 second. This of course is an explosion. This source of atomic energy is again blocked by the extreme scarcity of the isotopes of the heavy elements and the cost of the neutrons required for its production. It is perhaps a fortunate thing that no method has been found to release large amounts of atomic energy because such a reaction, once started, would be difficult to control and might result in a world-wide destruction of enormous extent. One should not shudder at the havoc, however, because its occurence is extremely improbable.

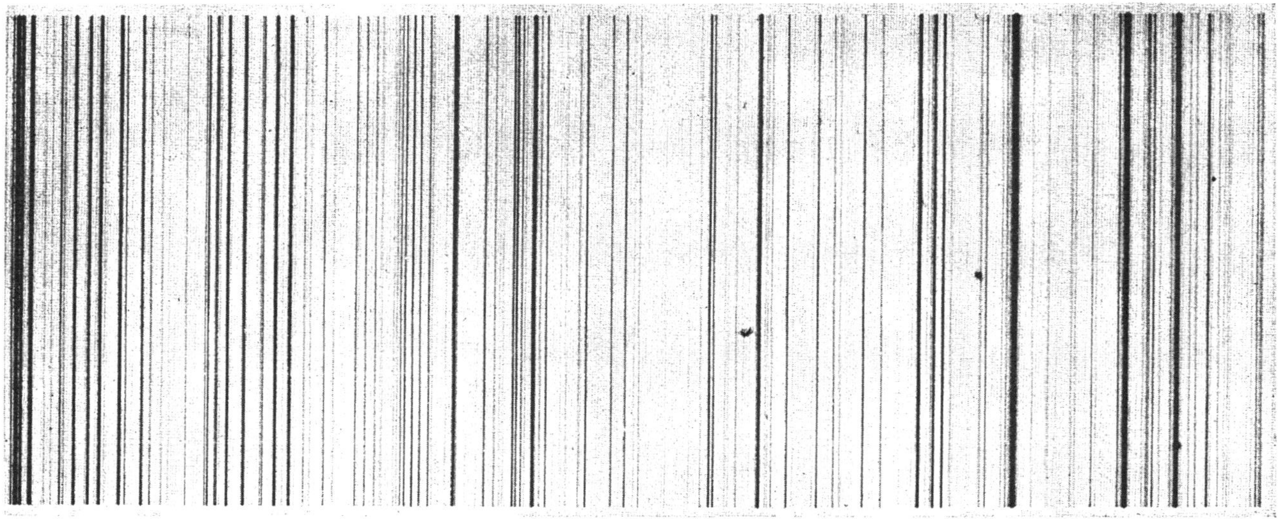

*Dean G. R. Harrison and associates at the Massachusetts Institute of Technology*

An enlargement of a detail of the spectrogram of one of the elements, showing the precision with which each of the lines may be identified.

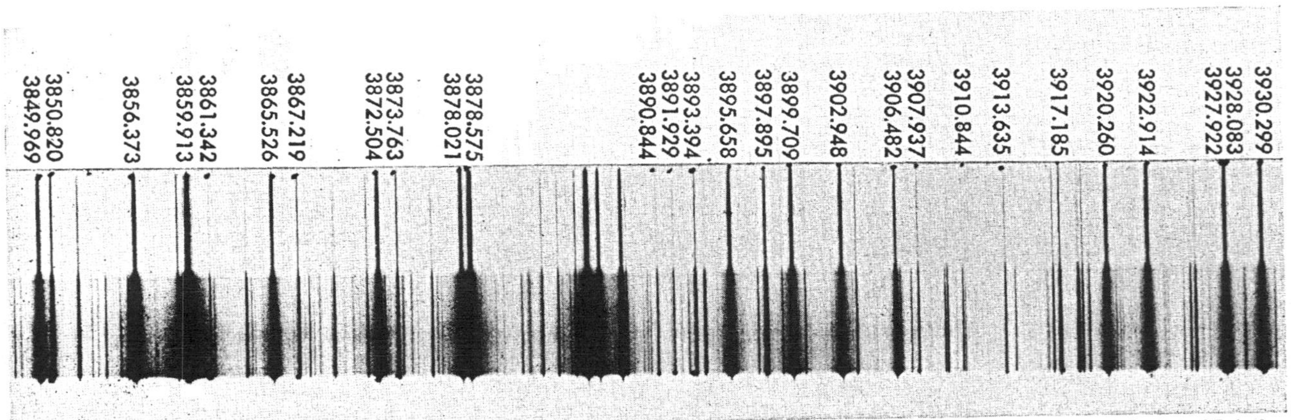


*Dean G. R. Harrison and associates at the Massachusetts Institute of Technology*

Part of the spectrogram of iron in the ultraviolet region between 3930.299 and 3849.969 angstroms. The corresponding frequency ranges, roughly, between $7.63 \times 10^{14}$ and $7.79 \times 10^{14}$ cycles per second. The photograph of the upper half of the spectrogram is exposed for a shorter time than the lower half so that sufficient space is left for the angstrom identification of the principal lines.

*Dr. E. O. Lawrence and associates at the University of California*

The world's largest operating cyclotron is located at the University of California. The magnet, on the left, weighs 220 tons and has a pole-face diameter of 60 inches. Accelerated deuterons with an energy of 16,000,000 electron-volts bombard the material to be transmuted in a storage chamber behind the magnet. It is operated largely for the production of radioactive isotopes employed in medical research. The cyclotron may be adjusted at close range only when not in operation. When active the operators must stand at a considerable distance behind a water barrier.

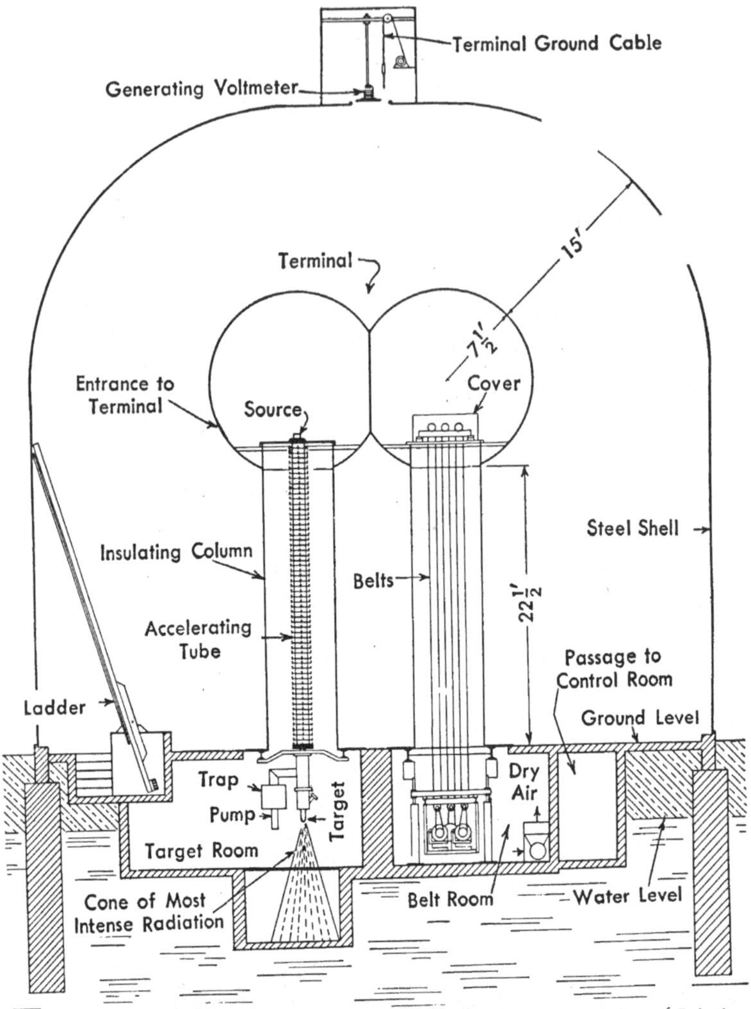


*Dr. R. J. Van de Graaff and associates at the Massachusetts Institute of Technology*

Cross-sectional diagram of the 2,500,000-volt Van de Graaff electrostatic generator.

CHAPTER II

# THE FLOW OF ELECTRICITY

## The Electric Current

Most of the useful applications of electricity are associated with a flow of positive and negative particles of electricity in opposite directions through space which may or may not be occupied by matter. This flow of electrical particles is called an "electric current," and its direction is always taken as the direction of the flow of the positively charged particles. The direction of a current is therefore always opposite to the flow of the negatively charged particles.

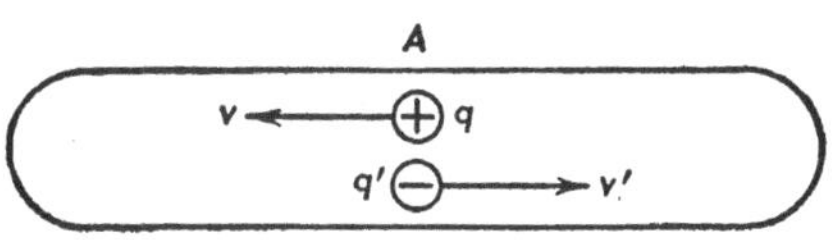


**Fig. 8.** The elements of an electric current

If at $A$ in Fig. 8, $q$ positive statcoulombs flow with velocity $v$ centimeters per second to the left, and $q'$ negative statcoulombs flow with velocity $v'$ centimeters per second to the right, the electric current (direction to the left) would be $qv + q'v'$ statcoulombs per second. One ampere equals 3 billion statcoulombs per second. A 60-watt, 120-volt lamp, for comparison, takes 0.5 ampere, or 1.5 billion statcoulombs per second. Since the positive charges in gaseous and metallic conductors are several thousand times as massive as the negative charges, which are electrons, $v$ is negligible compared to $v'$ so that the entire strength of the current may be attributed to the flow of electrons. Under these conditions an ampere would correspond to the flow of 6.25 million, million, million electrons per second.

## The States of Matter

The state or form of a substance depends upon its temperature and the surrounding pressure. At atmospheric pressure, mercury is a solid below − 38.7 degrees Centigrade, a liquid between − 38.7 and + 357.2 degrees, and a gas above + 357.2 degrees. For the proper understanding of vacuum-tube operation a distinction must also be made between a gas and a vapor. When mercury, for example, boils at atmospheric pressure at + 357.2 degrees, the resultant gas is usually called a "vapor" because a slight increase in pressure will return it to the liquid form.

A vapor is then a gas, which may be liquefied by a moderate increase in pressure. Hydrogen exists at atmospheric pressure as a solid below − 259 degrees, as a liquid between − 259 degrees and − 252.6 degrees, and as a gas above −252.6 degrees, but this gas is not called a vapor because it requires a very large increase in pressure to return it to the liquid form. Another easily observed distinction between the three states of matter is that a solid has a fixed volume and shape, a liquid has a fixed volume and a variable shape, and a gas or a vapor has a variable volume and shape.

### Conduction in Gases

Although not discussed in detail in the first chapter further mention must now be made of the fact that atoms of the elements, with some exceptions, combine with their own and other atoms to form molecules, which are the smallest particles of a substance to exist in a stable state. Helium, neon, argon, krypton, and xenon atoms combine neither with their own nor with other atoms so that their molecules are called "monatomic," and for these elements the terms "atom" and "molecule" are identical. Hydrogen, oxygen, and nitrogen and most other atoms combine with their own respective atoms to form a "diatomic" molecule; that is, two atoms to a molecule. They also combine with the atoms of many other elements and form various molecules of simple or complex nature.

It may be shown experimentally in several different ways that the number of molecules contained in a certain volume of gas at the same temperature and pressure is the same for all gases. At zero degrees Centigrade and atmospheric pressure the number is $4.44 \times 10^{20}$ molecules per cubic inch. At this temperature and pressure every cubic inch of air that we breath contains about $10^{20}$ molecules of oxygen and about $3.4 \times 10^{20}$ molecules of nitrogen. If you wish to see it spread out for the oxygen it is 100,000,000,-000,000,000,000 molecules per cubic inch.

Although the number of molecules in a given volume of gas is extremely large, they occupy only a small part of the total space and move about at a high velocity. This velocity is constantly altered in magnitude and direction by elastic impacts and rebounds with other molecules and with the sidewalls of the container. It is this impact of high-velocity molecules that causes the sides of a tank to bulge or even burst if the velocity is too high. At zero degrees Centigrade and atmospheric pressure the average (kinetic energy) velocity of the molecules of hydrogen gas is 1.17 miles per second; helium, 0.81 mile per second; nitrogen, 0.31 mile per second; oxygen, 0.29 mile per second. The muzzle velocity of a modern rifle bullet (0.51 mile per second) is less than half that of the hydrogen molecule. The molecules of hydrogen, moreover, collide about 10 billion times per second, and for the other gases mentioned above, about 5 billion times per second.

The molecules of a gas are not only in frequent collision with each other but also with rays and high-velocity particles which may enter the gas chamber from the outside. The rays may include $\gamma$-rays, X-rays, and cosmic rays, and the particles, electrons, protons, $\alpha$-particles, and neutrons. The impact of any of these rays or high-velocity particles may cause molecules in the gas to lose one or more of the orbital electrons surrounding its atoms. The molecule is then said to be "ionized"; that is, split into two charged parts, one a positive ion and the other a negative ion con-

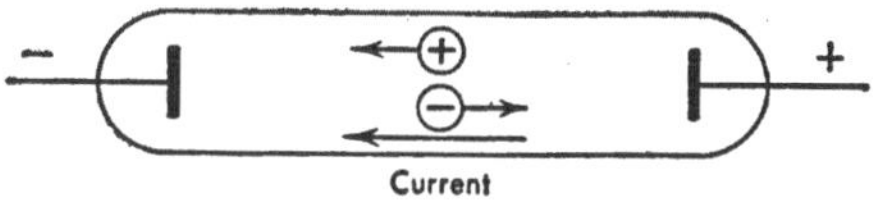

**Fig. 9.** Gaseous conduction

sisting of one or more electrons. A voltage applied to plates located at the ends of the gas chamber will cause the ions to move in opposite directions as shown in Fig. 9. An electric current will then be established in the gas with a direction from the positive to the negative plate.

If the voltage is moderate (not over 75 volts per inch between the plates), the current will not exceed $10^{-17}$ ampere per cubic inch of the gas. The current is limited to this saturation value by the number of impinging rays and particles of external origin which ionize the gas. The voltage may then be increased considerably without further increase in the current because all the existing ions are being pulled toward the plates. If the voltage is made many times 75 volts per inch, the current will begin to increase again because the ions themselves (principally the electrons) will attain sufficient velocity to ionize other molecules of gas by collision. After a very high voltage is reached ionization will become so general that a low-resistance arc will be formed between the terminals, and the gas is said to "break down."

There are many factors such as the volume and length of the gas chamber, and the nature and pressure of the gas, which influence the flow of electricity in gases, but the above brief description should give a general idea of the modern ionization theory. The magnitude of the pressure is particularly important because ionization due to collisions of internal origin under high pressure is inhibited by the short distances in which electrons may accelerate before colliding with a molecule. If insufficient velocity is attained, a collision will not result in ionization. For this reason a gas under high pressure may serve as an insulator.

### Conduction in Nonmetallic Liquids

There are $8.9 \times 10^{20}$ atoms in a cubic inch of hydrogen gas at zero degrees Centigrade and atmospheric pressure. When two atoms of hydrogen are combined with one atom of oxygen they form a molecule of water. One cubic inch of water contains about $1.6 \times 10^{24}$ atoms which is about 1800 times as many atoms as are contained in a cubic inch of hydrogen. It will be seen that in a liquid the molecules are more concentrated and cannot move about as freely as in a gas. They may only circulate in short orbits at comparatively low velocities. Unlike gases, liquids may be kept in an uncovered container although some of the molecules will still escape in the familiar phenomenon of evaporation.

Although the same illustration of the enormous atomic content of a given volume of matter might be applied to any of the states, it may best be demonstrated in the case of a liquid. For example, suppose we should tag each atom in an 8-ounce glass of water so that it could be recognized again wherever it might be found. We will pour this glass of water into any near-by river, lake, or ocean and wait (a very long time) until all the atoms have been uniformly diffused in all the waters of the earth. Then we will go anywhere on the earth, let us say to a small pool in the heart of Africa, and dip up a glass of water. How many of our original tagged atoms would we find? The answer is five thousand.

The molecules of a liquid being closer than in a gas are held together by the strong electrostatic forces between their charged particles. These forces are so great that the molecules are practically immune from disruption by the rays or high-velocity particles of external origin which do ionize the diffused molecules of a gas. For this reason some substances are less conductive in the liquid than in the gaseous state; pure water and oils, for example, are nearly perfect insulators in the liquid state but not in the gaseous state.

With certain exceptions most substances which dissolve readily in water are spontaneously dissociated into positive and negative ions. Although the cause of the ionization is not clearly understood it is nevertheless well established that in certain molecules, called "acids," the positive ion is always one or more protons. In other molecules, called "bases" or "alkalies," the negative ion is always one or more combinations of one atom of oxygen and one atom of hydrogen bearing one excess orbital electron. In a third type, called a "salt," the positive ion is always the atom of a metal which has lost one or more orbital electrons.

All of these naturally dissociated aqueous solutions are called "electrolytes." If a voltage is maintained between two plates or sheets of metal dipped in an electrolyte, the positive ions will flow toward the negative plate and the negative ions toward the positive plate, thus producing again an electric current. The velocity of these ions under ordinary conditions is extremely low; the hydrogen ion (a proton), for example, takes about 13 minutes to move 1 inch. Nevertheless an exceedingly large number of these ferryboats bearing opposite charges in opposite directions may produce a heavy electric current. In the storage battery of your automobile the current required to start the engine will range from 200 to 300 amperes.

The process of spontaneous ionization also takes place in some solids. A molecule of sodium chloride (table salt) contains a positive sodium ion (a sodium atom minus an electron) and a negative chlorine ion (a chlorine atom plus an electron). In this combination the molecule possesses essential dietetic properties although each of the un-ionized constituents is a deadly poison.

## Conduction in Solids

While the molecules in a gas move rapidly over zigzag courses and the molecules of a liquid move more slowly in restricted orbits, the molecules in a solid take up a fixed but vibrating position in a three-dimensional lattice. The number of molecules per cubic inch is only slightly greater than the number in a liquid state. Although spontaneous ionization does occur, as previously stated, in some solids, motion of the charged ions due to an applied voltage is negligible because the ions are tightly bound together by the associated electrostatic forces.

The nuclear attraction for the orbital electrons in the outer shells of the atoms of metals is not only small on account of the large intervening spaces but the resultant force on such electrons due to the combined attraction of all the adjacent nuclei becomes intermittently zero under the close packing existing in metals in either the solid or liquid state. The open spaces in metals thus become filled with free electrons which zigzag about at very high velocities in a manner similar to the molecules in a gas except that the average (kinetic energy) velocity of the electrons is much higher (about 70 miles per second). Another difference is that the free electrons in a metal are continually changing their identity. Some free electrons when passing close to an atomic nucleus are recaptured while others in the outer orbits of the atoms arrive at a position of neutralized forces and are set free.

Many substances, such as glass, porcelain, and rubber, possess the common properties of being chemically inactive, not being soluble in water, and having low density. Their molecules are spaced widely apart so that the orbital electrons in any molecule are not easily detached. In consequence these substances possess very few free electrons and are called "insulators."

If a voltage is applied between the opposite sides or ends of any metal containing free electrons, these electrons are pulled toward the positive side of the metal and produce an electric current. The molecule from which any electron is detached becomes positively charged and will be pulled toward the negative side of the metal but with negligible velocity by reason of its large mass and the restricted space in which it may move. The flow of electricity in metals is then interpreted as a flow of electrons from the negative to the positive terminal. Since these are negative particles the direction of the current, as previously stated, will be from the positive to the negative terminal of the metal.

The drift velocity of the free electrons in metals is very low; about 0.003 inch per second. If a powerhouse should start sending electric power to a factory 1 mile away on the first of January, the electrons leaving the powerhouse on that day would arrive at the factory in the third week of August. Since all the electrons in the circuit to and from the factory start moving at practically the same instant, electric power would be delivered to the factory in about five millionth of a second after the switch is closed in the powerhouse. The drift velocity of the individual electrons is extremely low but the propagation velocity along the connecting wires is only slightly less than the velocity of light.

## Conduction in a Vacuum

The word "vacuum" is used here to mean a space in which the pressure is less than atmospheric (14.7 pounds per square inch) and not a space in which the pressure is zero, because that condition is not attainable. The best vacuum obtained to date contains one ten-billionth of the original molecules which filled the space at atmospheric pressure. Expressed in another way a cubic inch of space in the most highly evacuated tube contains more than seven billion molecules, which is more than three times the number of people living in the whole world. It will be seen that conduction in a vacuum is actually the same as conduction in a low-pressure gas. Since low pressure means low density and greater space between the molecules, electrons released by ionization, thermionic, or photoelectric emission (described later) will reach high velocities because collision with other molecules is less frequent. When collisions do occur, the high velocity makes ionization of an intercepted molecule more certain.

A rarefied gas thus becomes a better conductor of electricity than the same gas at atmospheric pressure because there is more room for the electrons to attain high velocity. This principle does not continue indefinitely, however, because when the number of molecules per unit volume within the gas becomes extremely small the chance of collision is less probable. This change takes place gradually at a pressure of about one ten-thousandth of atmospheric pressure. A further decrease in pressure reduces the conductivity of the gas, which becomes zero at zero pressure. A very low-pressure gas thus becomes an excellent insulator, and it is possible that our future transmission lines may consist of highly evacuated pipes enclosing a central conductor which will conduct the outward current. The current will return through the pipe itself. It is possible that such systems may operate at 50,000,000 volts in contrast with our present open-wire maximum of 287,500 volts.

If the high-velocity electrons constituting the current come from the cathode and not by ionization of the low-pressure gas, conductivity in a vacuum takes on an entirely different aspect. The emission of electrons from the cathode may be established by heating the cathode or by bombarding it with photons, the carriers of visible and invisible light. These methods of electron liberation are described in detail in later chapters. If the electrons do originate at the cathode, more of them will reach the anode if nothing whatever blocks their transit through the intervening space. A perfect vacuum would thus become a good conductor of electrons if they are emitted by the cathode. Under these conditions the current would also be strengthened if the intervening space contained a few molecules; that is, contained a low-pressure gas. The high-velocity electrons originating in the cathode would then ionize some of the molecules by collision and cause the liberated electrons to increase the stream of electrons flowing toward the anode. The increased flow of electricity in some types of vacuum tubes depends upon this principle.

## CHAPTER III

# RADIO COMMUNICATION

### Thermionic Emission

At the beginning of the first chapter it was shown that electrons within the cold cathode of a cathode-ray tube may be drawn from its surface and pulled across the intervening space to the anode if a voltage of sufficient magnitude is maintained between the two terminals. This method, called "field emission," possesses little practical value with certain exceptions in sources of illumination because very few electrons are released and the rate of emission is not easily controlled. If the cathode, on the other hand, is heated directly or indirectly by an electric current, the free electrons within the cathode material will be given increased random velocities and many of them will escape to the surrounding space in a manner similar to the loss of molecules from a liquid undergoing evaporation. In this method, called "thermionic emission," electrons are emitted with little or no voltage between the terminals of the tube.

The temperature at which thermionic emission begins depends upon the nature of the cathode material. Tungsten is a good emitter at an operating temperature of 2400 degrees Centigrade, thoriated tungsten at 1900 degrees Centigrade, and barium and strontium oxide at 1000 degrees Centigrade. Vacuum tubes operating at low voltage usually have oxide-coated cathodes, but at high voltage tungsten or thoriated-tungsten cathodes must be used to withstand the consequent forces of disintegration.

### Construction and Operation of the Diode

The simplest application of thermionic emission is illustrated by the diode, shown in Fig. 10. A heated filament (the cathode) is centered in a hollow metal cylinder (the anode or plate) and all are located within a highly evacuated glass or metal tube with connecting pins in the base. When the heating current is alternating the cathode is usually an indirectly heated thimble placed over the filament.

Starting with zero voltage between the anode and the cathode the electrons emitted by the heated cathode form a dense cloud around the cathode and, by repulsion, prevent other electrons from escaping. This stoppage or reduction of emission by a spatial distribution of electrons is called the "space charge effect." With the application of a small voltage (positive on anode and negative on cathode) electrons are pulled from the cloud of electrons to the plate and a small current is established in the tube with a direction, as previously explained, from the anode to the cathode.

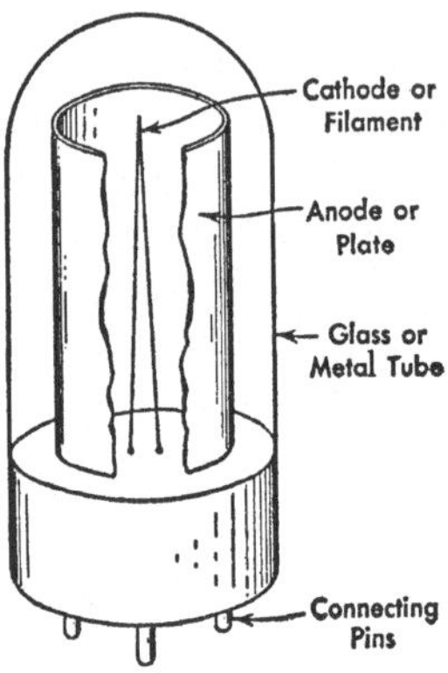


**Fig. 10.** Construction of the diode

When the voltage is sufficient to pull most of the electrons within the cloud toward the plate (point *A* in Fig. 11) a further increase in voltage causes the current to increase rapidly (*A* to *B*) because most of the repelling space charge has been removed and electrons are emitted in greater numbers from the cathode. The current finally reaches a fixed or saturated value at point *C* because the number of electrons emitted per second is limited by the temperature of the cathode. A re-

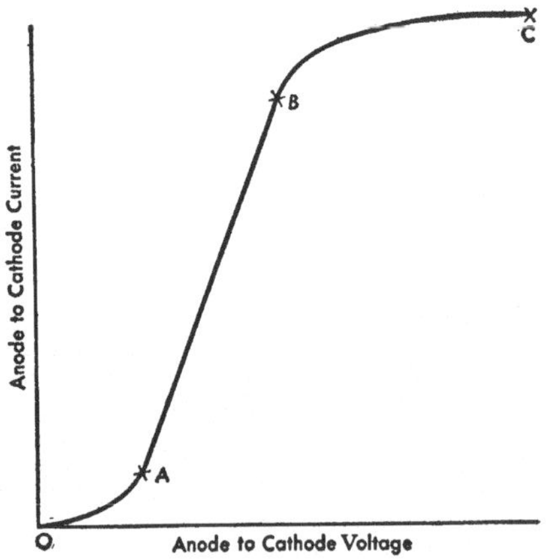


**Fig. 11.** Variation of the current with the voltage in a diode

versal of voltage within the tube will stop all outward motion of the electrons and reduce the current to zero. The diode may thus be employed as a rectifier of alternating current and will be discussed later in that capacity.

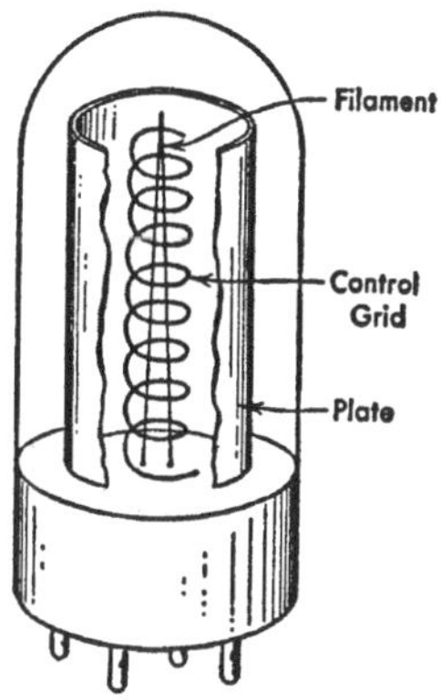


**Fig. 12.** Construction of the triode

## The Function of the Grid in the Triode

If a cylindrical screen of widely spaced small wires, called the "control grid," is placed between the anode and the cathode as shown in Fig. 12, the current established in the triode with a variable grid voltage and a fixed voltage from anode to cathode will vary as shown in Fig. 13. Since this controlling action of the grid voltage on the anode to cathode current constitutes one of the most important contributions to electronic development a great effort should be made to understand its significance. The chart, graph, or curve tells us that when the grid voltage is made *OA* volts (negative) with respect to the cathode,

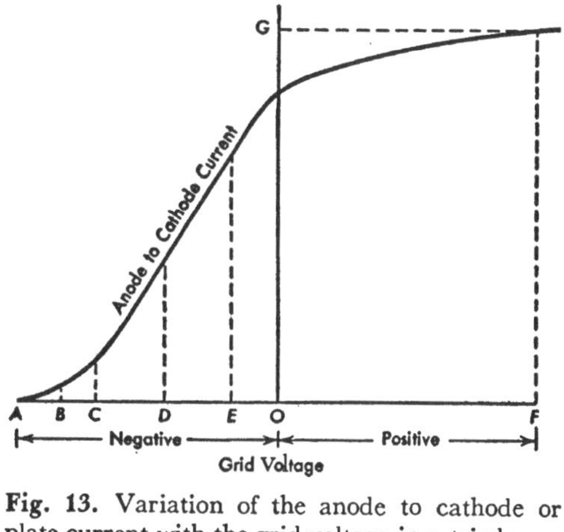


**Fig. 13.** Variation of the anode to cathode or plate current with the grid voltage in a triode

all electrons are pushed back to the cathode and no electrons flow within the tube. The anode to cathode, or plate, current is shut off.

If the grid voltage is decreased from *OA* to *OB* to *OC* to *OD* and *OE*, the fixed anode voltage steadily overcomes the reduced repelling action of the grid voltage and pulls electrons in increasing numbers toward the plate. A reversal and increase of the grid voltage to a positive value ($OF$) brings the anode to cathode current up to the saturation value ($OG$). It will be seen that a grid voltage originating in a weak source of energy may not only control the current supplied by a strong source of energy but may turn this current on and off millions of times per second. Before discussing the application of the triode to the transmission of signals, speech, and music, consideration must first be given to the radiation of energy itself. Later it will be seen that the triode is the indispensable element in such transmission.

## Radiation of Energy

If a long wire is set up perpendicular to the ground and a source of alternating voltage ⊙ is connected between the bottom of the wire and the ground ⏚, as shown in Fig. 14, an oscillating current will be established in the long wire and electromagnetic waves will be sent out in all directions at the velocity of light. This arrangement is called a "radiating" or "sending antenna." If a similar

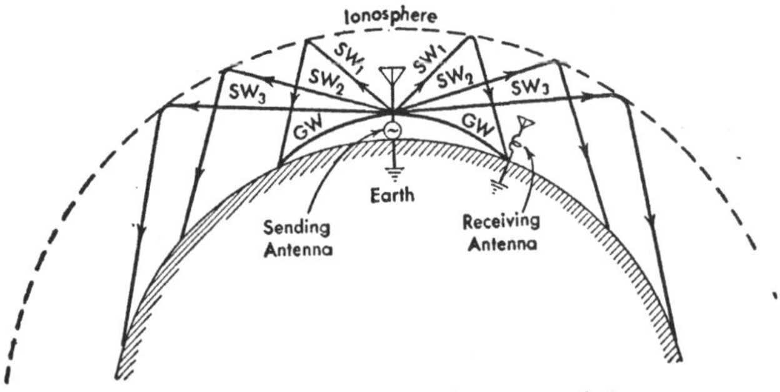


**Fig. 14.** Paths taken by ground waves and sky waves

"receiving" antenna is set up at any distance from the sending antenna, the receiving antenna will have a voltage induced in it by any electromagnetic wave that it intercepts, but not all the electromagnetic waves will arrive by the same path. Some of the waves, called "ground waves" ($GW$ in Fig. 14), will spread out horizontally and follow the curvature of the earth, while others, called "sky waves" ($SW_1$, $SW_2$, and $SW_3$ in Fig. 14) will take an upward path and be reflected back to the earth by the various shells of ionized gas, called "ionospheres," which surround the earth. At least six different ionospheres ranging in height from 50 to 250 miles above the earth have been identified at different times of the day and night or seasons of the year.

Both ground waves and sky waves are absorbed and weakened by passage over the earth and through the atmosphere. Sky waves are also weakened when reflected from the ionosphere. Only ground waves of the lower frequencies may be received reliably at any time of the day or year. Ground waves with frequencies ranging from 10 kilocycles per second to 550 kilocycles per second are employed principally for the transmission of signals, such as telegraph, rather than for broadcasting speech and music. Kilocycles per second will hereafter be designated as "kc" and megacycles per second as "mc." The broadcast frequencies for the general public range between 550 kc and 1550 kc and consist principally of ground waves. The lower the frequency of the ground wave the greater the distance it travels before absorption. Under normal conditions such waves are received at distances from 50 to 200 miles.

Sky waves of frequencies ranging from 1600 kc to 30 mc may be received by reflection under the best conditions at distances of 200 to 3000 miles. Reception, which is thought to be due to double or triple reflection between the earth and one of the ionospheres, has even been obtained at a distance of 12,000 miles, which is about half-way around the world. It will be noted from Fig. 14 that the sky waves may skip receiving antennas at certain locations near by and intercept antennas located at greater distances. Sky waves and ground waves taking different paths may also arrive at the same antenna, as shown by $GW$ and $SW_1$ in Fig. 14.

Reception by sky waves is usually best at night and in the winter, although certain frequencies are often received more clearly at other times. Since an ionosphere often changes its height very rapidly, reception by sky waves is subject to repeated weakening or "fading." This phenomenon, especially in the case of fluttering or rhythmic fading, may also be due to the arrival of two waves of the same frequency but periodically in opposition because they have come by paths of different lengths. Electromagnetic waves of frequencies exceeding 30 mc are not reflected to any degree by an ionosphere but penetrate it and disappear. They may be directed downward from the top of a mountain or a tall building, however, but the receiving antenna must be located within sight of the sending antenna with no obstruction between the two.

Although electromagnetic waves with a great range of frequencies pour in upon us from all sides we probably know less about their nature than any other physical phenomenon. All the energy that has come to us through the ages from the sun has been transmitted to us through 93 million miles of space by electromagnetic waves; also the light from twinkling stars, the planets, and the moon by reflection, from our own house and street lights, and from the telltale lines of the spectroscope. Even the firefly knows how to radiate electromagnetic waves.

At one time it was thought that such waves travel through an elastic substance, called "ether," which fills all space, even a perfect vacuum. The modern view modifies the ether concept and includes a train of particles of magnetic energy, called "photons," which move through space at the velocity of light and produce the same effect as the former electromagnetic waves in the ether. Although the photon theory steadily gains confirmation, it is still customary to visualize radiant energy in the form of electromagnetic waves because the effect in either case is the same.

## Reception of the Radiated Energy

Suppose a triode, two batteries, and headphones are connected at the bottom of a receiving antenna as shown in Fig. 15. The

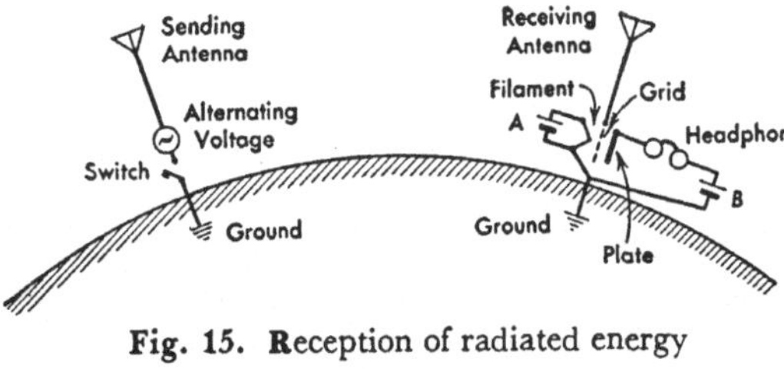


**Fig. 15.** Reception of radiated energy

A battery (symbol ▮|) heats the filament or cathode (symbol ⌒) The antenna is connected to the grid (symbol ---). The negative terminal of the B battery (same symbol as A) is connected to the ground (symbol ⏚) and the positive terminal is connected through the headphones (symbol ⌒O) to the anode or plate (symbol ——) of the triode. Every time the switch (symbol —⁄—) in the sending antenna is closed, the source of alternating voltage (symbol ⊝)

establishes an oscillating current in the sending antenna which sends out electromagnetic waves and induces a small alternating voltage in the receiving antenna.

Since this induced voltage is impressed upon the grid of the triode and will alternate from $OF$ to $OA$, as shown in Fig. 16 *a*, the current in the headphones will pulsate from zero to $OG$, as shown in Fig. 16 *b*, but will not reverse or alternate because the triode rectifies the current as shown in Fig. 13. The diaphragm in the

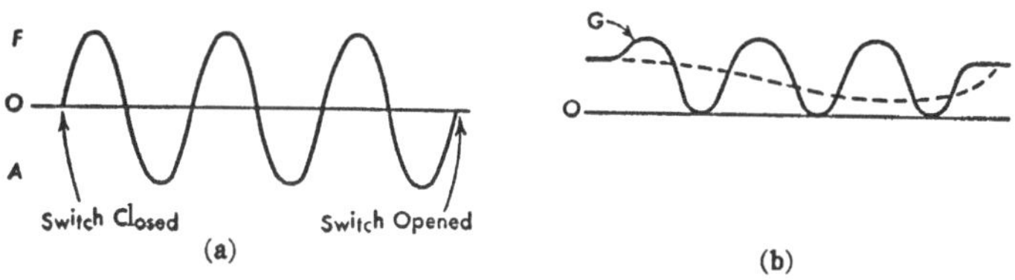


**Fig. 16.** **a.** Variation of the grid voltage at the receiving station. **b.** Variation of the current and the motion of the diaphragm in the headphones

headphones will not vibrate in response to the high-frequency pulsating current because it is too massive to follow such high-frequeney impulses. The diaphragm, following the dotted line in Fig. 16 *b*, will be bent once or click every time the switch is closed in the sending antenna.

If an "interrupter" is connected between the switch and the source of alternating voltage in the sending antenna so that the circuit is opened and closed, say 100 times per second, every time the switch is closed, the diaphragm in the headphones will vibrate 100 times per second and produce an audible hum or buzz of that frequency. Telegraphic messages in dot-and-dash code may thus be sent over a considerable distance from the sending antenna to the receiver. The distance will depend upon the strength and frequency of the current in the sending antenna and the sensitivity of the receiver.

## Amplitude Modulation

Let us now substitute a microphone for the switch and interrupter in the sending antenna. A microphone contains a diaphragm which vibrates accurately in response to the pressure and frequency of sound waves which impinge upon it and by various methods of construction weakens or strengthens the sending antenna current in direct proportion to the pressure and frequency of the sound waves. We may also increase the sensitivity of the receiver by making the changes and additions shown in Fig. 17. One coil (the primary) of two closely wound coils of wire is now connected between the receiving antenna and the ground. The other coil (the secondary) together with a C battery is connected between the grid and the filament.

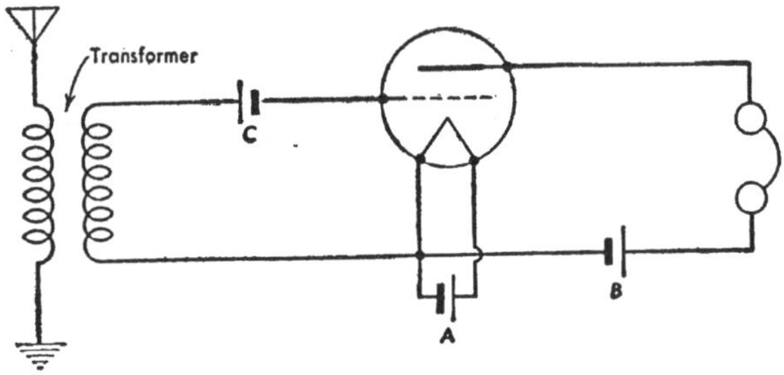


**Fig. 17.** A simple, untuned, one-tube receiving set

The two closely wound coils constitute a "transformer." Alternating current in the primary winding will induce an alternating voltage in the secondary winding just as an alternating current in a sending antenna induces an alternating voltage in a receiving antenna. If the secondary winding contains many more turns than the primary winding the voltage induced in the secondary winding may be many times greater than the voltage induced in the receiving antenna. The C battery with its negative terminal connected to the grid and with its voltage adjusted to, say, *OB* volts in Fig. 13 (page 31) will keep the plate current at a low constant value when there is no induced voltage in the antenna.

Suppose the microphone at the sending station "modulates" the high-frequency current in the sending antenna from that shown in Fig. 18 *a* to that shown in Fig. 18 *b*. The voltage impressed upon the grid at the receiving station will then vary as shown in Fig. 18 *c*, the voltage wave being transferred to the negative side by the C battery shown in Fig. 17 and the letters *O*, *C*, *B*, *A* referring to Fig. 13. The plate current, referring again to Fig. 13 (page 31), cannot reverse but will vary as shown in Fig. 18 *d*. The diaphragm in the headphones on account of its inertia will not respond to the original high-frequency current but will follow the average of the

*Zenith Radio Corporation*

The turnstile antenna of broadcasting station WWZR radiates a frequency-modulated wave of 45.1 megacycles per second frequency from the top of the Field Building in Chicago. This type of antenna with four bays and folded half-wave length arms provides dependable reception at a distance of at least 100 miles.

*The Yankee Network*

The turnstile antenna of the frequency-modulation station WMTW located on the summit of Mount Washington in New Hampshire. The frequency is 43.9 megacycles per second and the top of the antenna is 6394 feet above sea level. Neither ice nor snow obstructs the radiation.

*National Broadcasting Company*

Television antenna on the Empire State Building in New York City. The sound signals are broadcast from the ring with gaps at the top and the television or video signals from the cigar-shaped projections at the bottom.

*Westinghouse Electric and Manufacturing Company*

Two of the four air-cooled, 50,000-watt, modulating tubes at broadcasting station KDKA are spares. By pushing a button one pair can be replaced by the other without appreciable interruption in transmission.

rectified current as shown by the dotted line in Fig. 18 *d*. The headphones will thus respond to the tops of one side of the original modulated alternating current in the sending antenna, shown in

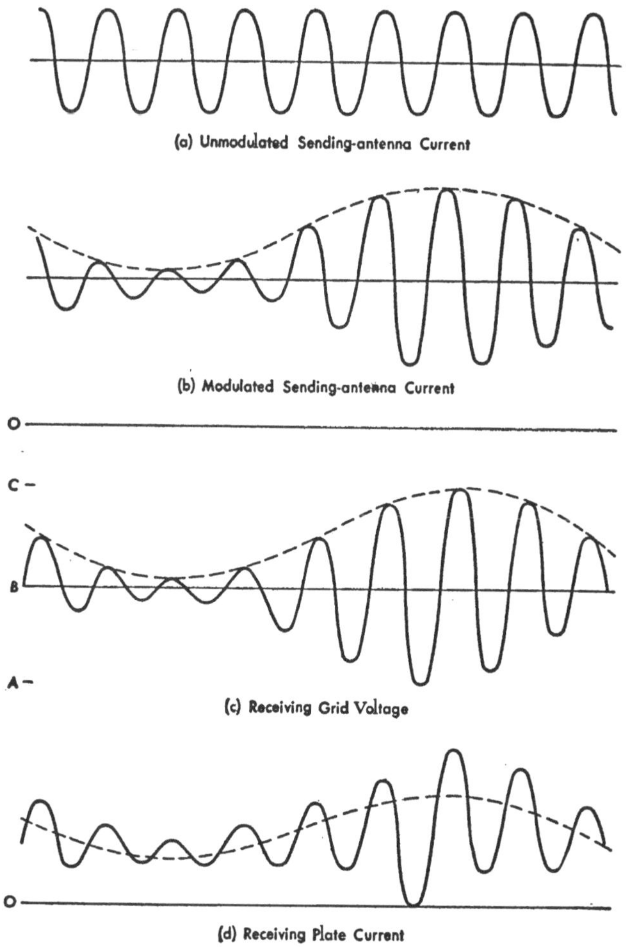


**Fig. 18.**

Fig. 18 *b* and thereby reproduce the original sound waves impressed upon the microphone in the sending station.

In this application of the triode as a rectifier or detector, the plate current is not only maintained in the same direction but is amplified in that direction because a change of grid voltage (Fig.13) from *OA* to *OB* produces less change in the plate current than a change of grid voltage from *OB* to *OC*. A diode will serve equally well as a rectifier or detector but will give no amplification of the signal.

## More Amplification

The elementary receiving set shown in Fig. 17 may supply sufficient plate current to operate sensitive headphones but not enough to operate a loud-speaker. This insufficiency may be overcome by substituting the primary winding of another transformer for the headphones of Fig. 17 as shown in Fig. 19. The A battery, although still required, is hereafter omitted to simplify the diagram.

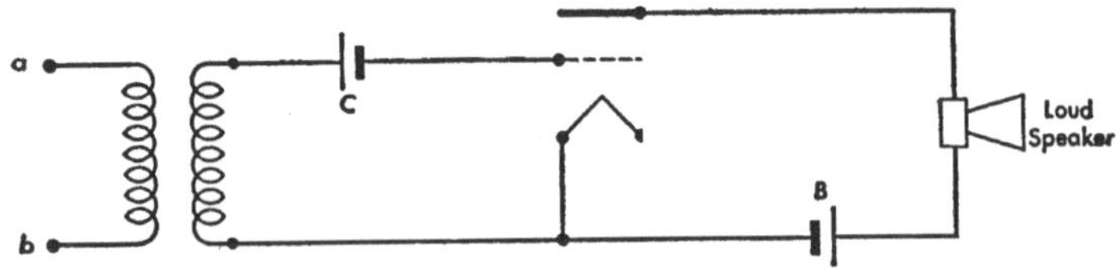


**Fig. 19.** One-stage amplifier with transformer coupling

The terminals, *a* and *b*, of the new primary winding represent the former terminals of the headphones. The voltage formerly impressed across the headphones is again increased by the transformer principle already described and the higher voltage of the secondary winding is impressed upon the grid of the second triode. Two stages of amplification are usually required for a loud-speaker. The interconnection between triodes in an amplifier circuit is called a "coupling." Although a "transformer" coupling is shown in Fig. 19 for the sake of simplicity, other couplings using resistors, inductors, and capacitors (these terms are explained later) are often employed. Transformer couplings are used principally in radio-frequency amplifiers and resistor-capacitor couplings in audio-frequency amplifiers.

The negative grid bias may be adjusted in Fig. 19 so that a change of grid voltage from *OC* to *OE* in Fig. 13, page 31, on the steep part of the curve, will produce a much greater increase in the plate current than the former change from *OA* to *OC*. One part of the "grid voltage versus plate current" curve in Fig. 13 may thus be applied effectively for detection purposes and another part for amplification.

## Tetrodes and Pentodes

While the triode may be utilized as described for the detection and amplification of radio-frequency and audio-frequency currents or voltages, it nevertheless has two defects which may be corrected to a considerable degree by adding two more grids, called the "screen" grid (*SG*), and the "suppressor" grid (*RG*), as shown in Fig. 20. The original single grid in the triode is called the "control" grid (*CG*).

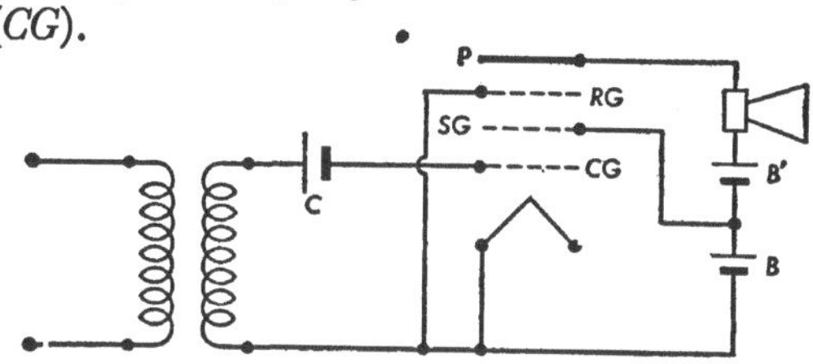


**Fig. 20.** The construction of a pentode

In the triode construction shown, for example in Fig. 19, the varying voltage of the plate induces another voltage in the near-by control grid in the same manner that a sending antenna induces a voltage in a receiving antenna. This induced voltage in the control grid will not only distort the voltage supplied to the grid by the transformer secondary and reduce the fidelity of reproduction but may also set the tube into oscillation (described in detail later) and produce an undesirable whistle or hum in the loud-speaker.

If a screen grid is connected to a point in the wire connecting the B and B′ batteries in Fig. 20, it will be maintained at the constant voltage of the B battery and screen the control grid from the fluctuating voltage of the plate. This two-grid vacuum tube is called a "tetrode" because it contains four elements; the filament, the control grid, the screen grid, and the plate.

In the other defect of the triode, and also of the tetrode, high-velocity electrons which bombard the plate will usually detach other electrons from the plate in a process called "secondary emission." If these detached electrons strike the grid, its voltage will again be distorted. A third grid, called the "suppressor" grid, placed near and often surrounding the plate, will catch most of these detached electrons and return them to the filament through a direct connection within the tube. The three-grid tube is called a "pentode" because it contains five elements.

## Tuning in

In the foregoing discussion it has been shown how signals, speech, or music may be transmitted through space from one place to another. If there were but one sending station there would be little else to add. Since there are actually hundreds of sending stations in operation at different carrier frequencies at the same time, every receiving set must have some means for excluding all but one station.

Every electric circuit possesses three distinct properties, or parameters, called "resistance" ($R$), "inductance" ($L$), and "capacitance" ($C$). A long thin wire, called a "resistor," made of a poor conducting material has a high resistance while a short, large wire made of a good conducting material has a low resistance. Any configuration or coil of wire, called an "inductor," in which a given current causes a large magnetic flux to surround many turns of wire, has a high inductance, and a low inductance if a given current causes a small magnetic flux to surround a few turns of wire.

An assembly of two conducting materials separated by an insulating material is called a "condenser" or "capacitor." The capacitance of a capacitor depends directly upon the area of the adjacent conducting surfaces and inversely as the distance between these surfaces; also upon a property of the insulating material, called the "dielectric constant." A capacitor has a high capacitance when its conducting surfaces have a large area, are located closely together, and are separated by an insulator with a high dielectric constant. Conversely the capacitance will be low for small areas spaced far apart and separated by an insulating material with a low dielectric constant. Resistances are measured in "ohms," inductances in "henrys," and capacitances in "farads."

The strength of an alternating current is always designated by its direct-current equivalent with the same heating value. This is called the "effective" value and for a sinusoidal current or voltage equals the maximum value divided by the square root of 2 (1.414). When an alternating voltage of $V$ volts is impressed upon a resistor of $R$ ohms resistance the current ($I$) in amperes equals $V/R$. If the same voltage is impressed upon an inductor of $L$ henrys inductance and negligible resistance the current equals $V/2\pi fL$, where $f$ is the frequency in cycles per second. The same voltage impressed upon a capacitor produces a current equal to $V 2\pi fC$. The mathematics involved in the above statements includes only simple multiplication and division.

It will be noted that an increase of frequency reduces the current in inductors and increases the current in capacitors. The other important point is that inductors and capacitors operate at all times in opposition; that is, when the inductor opposes the current at any instant the capacitor aids it, and vice versa. If we then connect just the right amount of inductance in tandem or series with a capacitance, they will cancel each other and the current will equal $V/R$. This cancellation, called "resonance," will occur when $2\pi fL = 1/2\pi fC$ or when $f = 1/2\pi\sqrt{LC}$. In the process of "tuning in" the variable capacitance, $C$, in Fig. 21, is adjusted until the circuit is in resonance at the frequeney of the desired signal. All other signals of different frequencies will be suppressed — not eliminated but made inaudible in the headphones. The response of the headphones or loud-speaker will be greater if the circuit is not exactly in resonance. The current in the headphones (the plate current in Fig. 13) is greatest when the grid voltage is a maximum positive. The grid voltage in Fig. 21 is the same as the voltage across the variable capacitor, which is $I/2\pi fC$. This voltage may be made considerably greater than at resonance by reducing $C$. $I$ will decrease slightly but the quotient will be greater than before. "Tuning in," then, consists of an adjustment of the variable capacitor to a capacitance slightly less than the resonance value for the desired frequency.

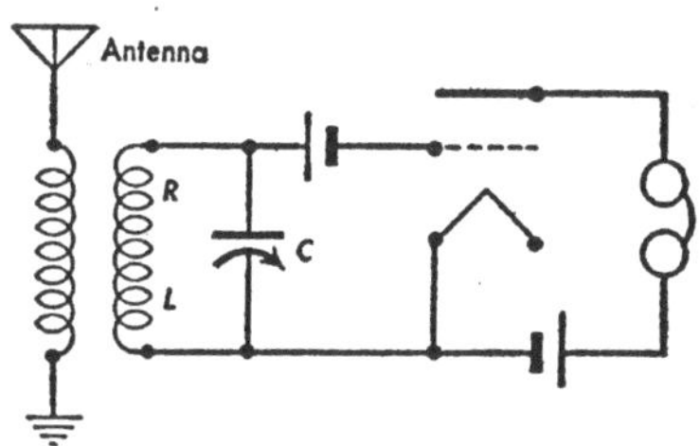

**Fig. 21.** The tuning circuit

## Oscillators

The source of alternating voltage at a sending station must generate voltages at frequencies ranging for different stations and different purposes from 10 kc to 300 mc or higher. The speed in revolutions per minute of a rotating generator of the type used to supply alternating current for power and light equals 120 times the frequency divided by the number of north and south poles on the generator. From this relationship it will be seen that for the lowest radio frequency (10 kc) a generator with 10 poles must run at 120,000 revolutions per minute and for a frequency of 300 mc at 3.6 billion revolutions per minute. Since any generator operated at such speeds would fly apart, other means must be found for generating the high-frequency voltages required for radio communication.

High frequencies may be generated quite easily in various types of vacuum-tube circuits, called "oscillators." A simple type of oscillator is shown in Fig. 22. The variable capacitance, *C*, is first adjusted to make the *R–L–C* circuit resonant at the desired frequency. A small variation of the grid voltage due to any cause will produce a slight pulse of plate current which now passes through a coil coupled to the *R–L–C* circuit and with an inductance of *L′* henrys. A pulse of current in *L′* will induce a small voltage in the *R–L–C* circuit which will again vary the grid voltage.

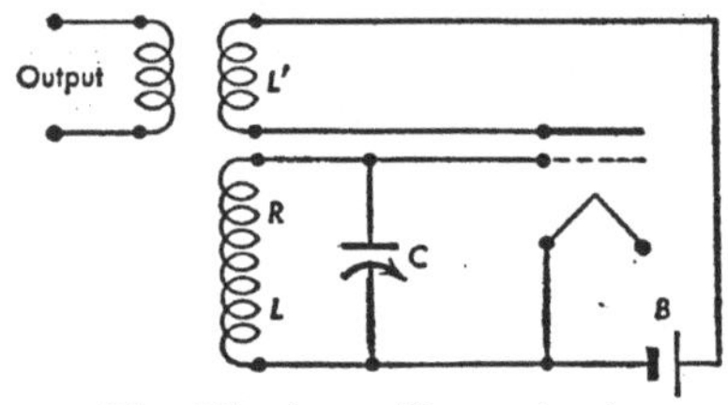

**Fig. 22.** An oscillator circuit

By this process of "regeneration" the alternating current in the *R–L–C* circuit will build up to a considerable magnitude at the resonant frequency. The high-frequency current in *L′* will also induce a voltage of the same frequency in the secondary winding of the output transformer. The source of this high-frequency power is the B battery, thus illustrating how a direct-current source of power, through inductors, capacitors, and vacuum tubes, may deliver alternating-current power at any desired frequency.

Oscillators are also included in the most efficient types of receiving sets. In the superheterodyne receiving set the modulated high-frequency voltage from the receiving antenna is amplified and then impressed together with an unmodulated (fixed-frequency) oscillator voltage of lower frequency on the first detector circuit. The resultant voltage impressed on this circuit is a modulated voltage with a comparatively low intermediate frequency equal to the difference of the two impressed frequencies. This intermediate frequency is maintained constant for any frequency of tuning. In other types of receiving sets called "tuned radio-frequency" sets the circuit connected to the first detector tube must be designed to respond to all frequencies to which the set may be tuned. In the superheterodyne receiver the detector tube circuit may be designed to respond more accurately and efficiently because it must be adjusted only for a fixed intermediate frequency.

## Frequency Modulation

The principal objection to the amplitude-modulation system is the propensity of a receiver to pick up and amplify high-frequency voltages due to amplitude-modulated electromagnetic waves coming from any source such as lightning, aurora borealis, X-ray machines, sparking commutators on motors, and the intermittent contact of a high-voltage wire with the branches of a tree. All of these sources produce the undesirable noises, called "static," which often ruin the quality of reception in the usual type of receiver. The extraneous noises will sometimes be loud enough to obscure the desired speech or music. All efforts to remove static from amplitude-modulation systems have been failures.

In a frequency-modulation, hereafter called "FM," sending station a high-frequency oscillating current of definite frequency is established in the antenna in the same manner as that established

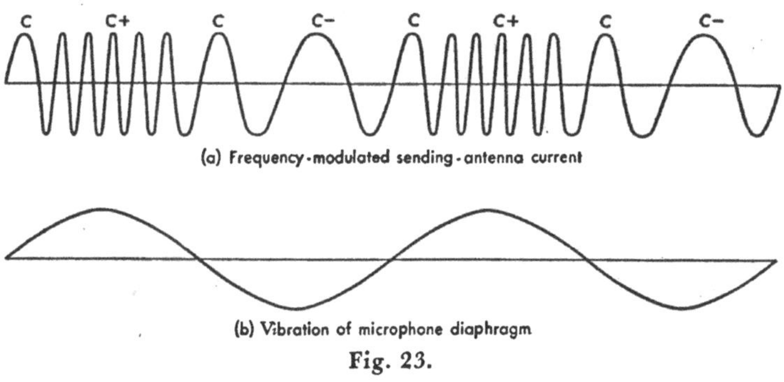


(a) Frequency-modulated sending-antenna current

(b) Vibration of microphone diaphragm

Fig. 23.

in the amplitude-modulation, hereafter called "AM," sending station. Instead of superposing an audible wave on the carrier wave, as shown in Fig. 18 *b*, for the AM system, the microphone in the FM system changes the frequency as shown in Fig. 23 *a*. The vibration of the microphone diaphragm is shown in Fig. 23 *b*. The unmodulated antenna current is shown at *C*, *C*, *C*, and *C*, the modulated current with increased frequency at $C+$ and $C+$, and the modulated current with decreased frequency at $C-$ and $C-$. The frequency of the audible wave or the vibration of the microphone diaphragm equals the number of $C+$ and $C-$ pairs per second. The loudness or amplitude of the audible wave equals the change in frequency; that is, $C+$ minus $C$, or $C$ minus $C-$.

In the FM sending station the microphone varies the voltage impressed on the grids of triodes located in the carrier-wave oscillator circuit and changes their capacitances. Since the frequency of the current induced in an oscillator circuit varies with the capacitances of its elements, as previously described, the frequency of the antenna current and the consequent carrier wave will change in accordance with the varying vibration of the microphone diaphragm.

In an FM receiver the antenna voltage, varying also as shown in Fig. 23 *a*, is detected or demodulated by a frequency "discriminator" which usually contains some inductors, capacitors, and two diodes. The discriminator is adjusted so that the audio-frequency output current is zero when the frequency of the input voltage equals that of the unmodulated carrier wave; that is, at $C$, $C$, $C$, and $C$ in Fig. 23 *a*. A slight increase in that frequency raises the plate voltage of one diode and a slight decrease raises the plate voltage of the other diode. This reversing voltage, which is applied to the audio-frequency amplifier and then the loud-speaker, reproduces the original audio wave impressed upon the microphone at the FM sending station.

It is often asked how an FM receiver may be tuned when the frequency of the voltage induced in the receiving antenna is a variable. Since the change of frequency of the carrier wave is very small, the frequency of the induced voltage varies very little and the tuning circuit may be designed to give loose tuning; that is, over a range sufficient to remain tuned to the expected changes in frequency. For this reason FM receivers cannot be tuned as sharply as AM receivers, but this is not a serious fault unless a great many FM transmitters of about the same frequency and power should supply the same area.

### Photoelectric Emission

In the previous chapter it was shown that certain metals and oxides emit electrons when heated. From this phenomenon of thermionic emission various vacuum tubes were developed which proved to be indispensable in the radio communication of sound and many other applications. Some metals, particularly the "alkali" metals (lithium, sodium, potassium, rubidium, and cesium), also emit electrons when their surfaces receive a high-frequency radiation such as ordinary light. This type of emission, which also applies in some degree to the "alkaline earth" metals (beryllium, magnesium, calcium, strontium, and barium), is called "photoelectric" emission.

The most important factor controlling the photoelectric emission from a given material is the frequency or wave length of the impinging radiation. Although the frequency, $f$, in cycles per second will be used throughout the following discussion it may be stated here that wave length, $\lambda$, in centimeters equals $3 \times 10^{10}/f$. Other units of wave length are the "angstrom" ($10^{-8}$ centimeter) and the "micron" ($10^{-4}$ centimeter). The energy imparted to an electron emitted photoelectrically equals $6.55 \times 10^{-27}f$ ergs minus the energy required to bring the electron from the surface of the metal. An electron emitted directly from the surface will bear the full amount. As previously stated photoelectric emission is produced only by high-frequency radiation. Most materials give no response to a radiation below $4 \times 10^{14}$ cycles per second (red light), which is also the lower limit for human vision. Many materials respond only to radiation frequencies in excess of $8 \times 10^{14}$ cycles per second (violet light), which is the upper limit of human vision.

Photoelectric emission is thus fundamentally a reaction to the red, orange, yellow, green, blue, and violet frequencies that affect the human eye and to some degree to the higher or ultraviolet frequencies. The frequency at which photoelectric emission begins for any material is called the "threshold" frequency for that material. Very few materials have a threshold frequency below $4 \times 10^{14}$ cycles per second; that is, in the infrared frequencies. This is a serious limitation because most of the radiation from an incandescent lamp, for example, is in the infrared frequencies. Fortunately certain oxides of metals do respond to the high values of infrared frequencies, notably a thin layer of cesium oxide on silver, which responds to frequencies as low as $2.5 \times 10^{14}$ cycles per second. The energy imparted to each emitted electron varies directly with the frequency of the radiation but the number of electrons emitted varies directly with the intensity of the radiation. No electrons will be emitted by a radiation frequency below the threshold value of any material no matter how great the radiation intensity may be.

## The Phototube

A photoelectric cell or "phototube" is usually constructed as shown in Fig. 24. The cathode is a half-cylindrical sheet of metal covered on the inner side with a thin layer of the photoemissive material (often cesium oxide on silver), while the anode is a straight coaxial wire of low photoemissivity. The enveloping tube is made of glass for visible or infrared frequencies but must have a quartz or special glass window for the higher (ultraviolet) frequencies which are absorbed by ordinary glass. For different applications the tube is either highly evacuated (vacuum type), or filled (gas type) with an inert gas, such as argon, at a pressure less than atmospheric.

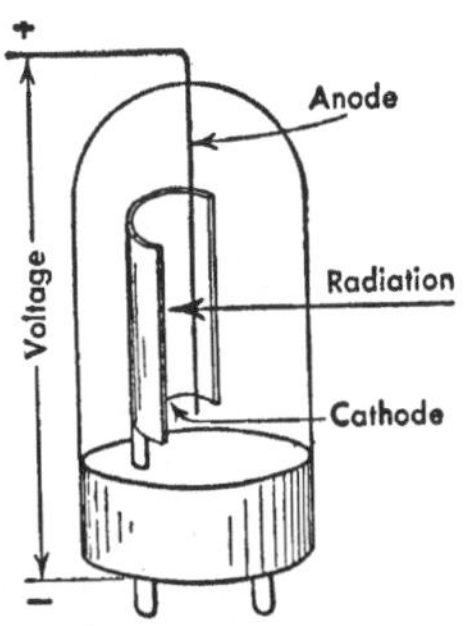

**Fig. 24.** Construction of a phototube

With the anode made positive the performance of each type of phototube under a definite radiation above the threshold value is shown in Fig. 25. It will be noted that the anode current in the vacuum type of phototube remains constant when the anode voltage has reached about 20 volts. Since ionization with this type of phototube is negligible, the anode current consists only of electrons drawn from the cathode and varies only with the intensity of the radiation. This phototube is best adapted to applications requiring exactly the same response at all times to the same radiation,

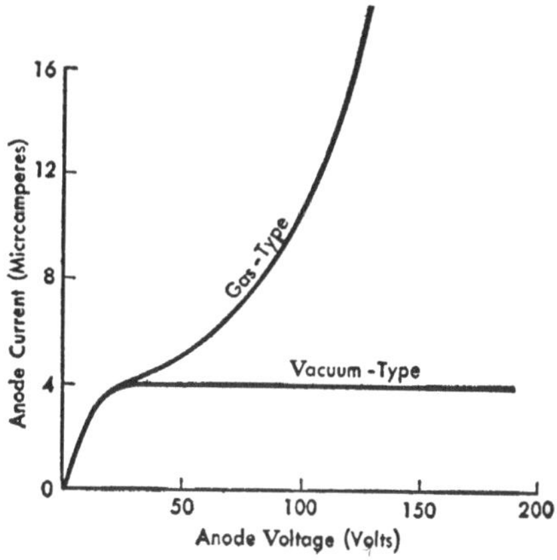


**Fig. 25.** Variation of current with voltage in a phototube under constant radiation

quick response to very rapid changes in the radiation, or in circuits in which the voltage impressed upon the tube might reach high values.

The gas-type phototube, at anode voltages above 20 vol s, gives a much stronger anode current than the vacuum type because the current is increased by the ionization of the argon gas. Since the anode current is a combination of the photoemissive current and the ionization current, its response to varying radiation is less constant and slower than in the vacuum type. The anode current in either type will increase approximately in direct proportion to the intensity of the radiation.

For this reason the gas type must contain externally some current-limiting device because a large increase in radiation might produce an anode current sufficiently large to burn out the tube. The gas-type phototube responds with sufficient fidelity and speed for most purposes and is used for most phototube applications because it delivers a stronger current than the vacuum type.

## Proof of the Quantum Theory

Before discussing various important applications of phototubes a simple demonstration will now be given of the reality of the quantum theory. A source of light of one candle luminous intensity (a 60-watt incandescent lamp, for example, gives about 60 candles) and with a single radiation frequency of $5.4 \times 10^{14}$ cycles per second (yellowish-green light) is known experimentally to radiate $2 \times 10^5$ ergs per second. Suppose this lamp is placed at the center (*A* in Fig. 26) of a hollow sphere 300 centimeters in radius. The area of the interior surface of the sphere would be $10^6$ square centimeters and the energy received by each square centimeter from the lamp would be $2 \times 10^5/10^6$ or 0.2 erg per square centimeter per

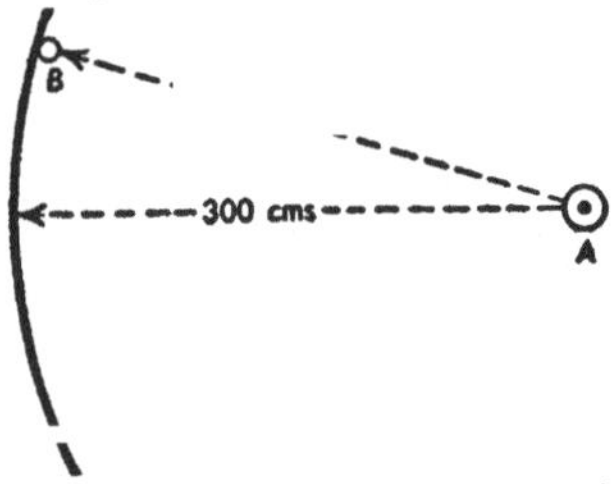


**Fig. 26.** A source of light at the center of a sphere sprays the inside of the sphere with drops of energy, called "quanta"

second. An atom at *B* (about $10^{-15}$ square centimeter in area) would then receive $0.2 \times 10^{-15}$ or $2 \times 10^{-16}$ erg per second. At the stated frequency an electron emitted from the surface of this atom by the radiation from the lamp would possess kinetic energy equal to $6.55 \times 10^{-27} \times 5.4 \times 10^{14}$ or $3.5 \times 10^{-12}$ erg. It will be noted that the energy received by the ejected electron is $3.5 \times 10^{-12}/2 \times 10^{-16}$ or 17,500 times the amount of energy absorbed by the whole atom in one second if the energy radiated by the lamp were uniformly distributed over the entire inside of the sphere.

With uniform distribution it would take 17,500/3600 or about 5 hours for an atom to receive enough energy to eject an electron bearing that energy. Since electrons are ejected the instant the lamp is turned on, it is evident that the lamp must radiate pulses or quanta of energy and each quantum is fixed by the frequency of the radiation. The lamp thus sprays the inside of the sphere with pulses of energy, or photons, which take different successive paths so that each atom will intercept a photon only once in about 5 hours. This is perhaps the simplest proof that energy is not radiated steadily and uniformly in all directions but is radiated intermittently in fixed quanta in changing directions.

## Sound Reproduction on Film

In the sound-track type of reproduction the sound is recorded on the film, (1) by a track of constant width and varying density

(Movietone), or (2) by a track of constant density and varying width (Photophone). The sound track for the Movietone film is photographed on a separate film simultaneously with the filming

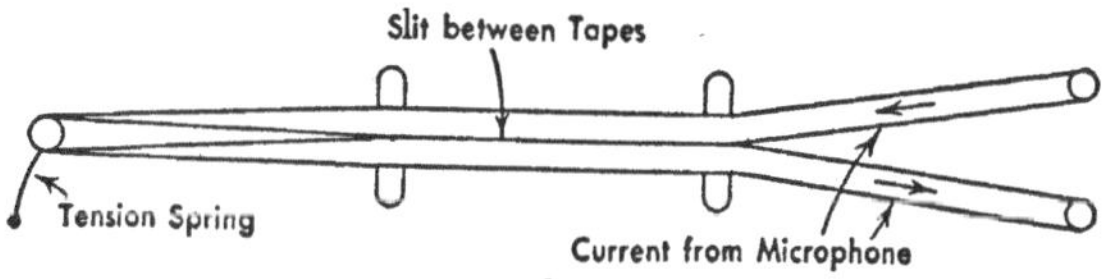


**Fig. 27.** A light valve in which the distance between two tapes varies with the current

of the motion picture. Microphones pick up the sound and deliver their amplified current to a light valve located within the sound-recording machine. The light valve consists of a double strip of duralumin tape which is stretched over two bridges as shown in Fig. 27. The light valve is located between the poles of an electro-

**Fig. 28.** Assembly of light valve, source of light, electromagnet, and film

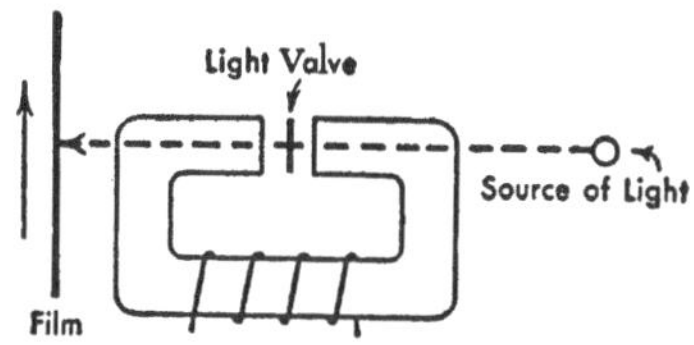


magnet as shown in Fig. 28. The direction of the magnetic flux is perpendicular to the plane of the double strip.

A narrow slit, centered on the contact line of the double strip, extends through the core of the electromagnet from the source of light to the film. When the strips of the light valve are separated, light passes through the slit to the recording film. Since the current in the double strip flows in opposite directions in each strip, the reaction of this current with the magnetic flux will cause the strips to separate and the distance between the inner edges of the strips will be proportional to the current flowing in the double strip and is therefore proportional to the amplitude of the sound wave picked up by the microphones. The number of vibrations (open and close) of the light valve per second will correspond to the various frequencies of the original sound wave.

A continuous series of parallel lines of varying thickness and number per unit length will be photographed on the moving film. Since the film moves at a speed of 90 feet per minute (18 inches per second), a musical note of 256 vibrations per second (middle C) will produce 14.2 parallel lines per inch on the film. Each line is 0.1 inch in width and its thickness is proportional to the intensity of the sound.

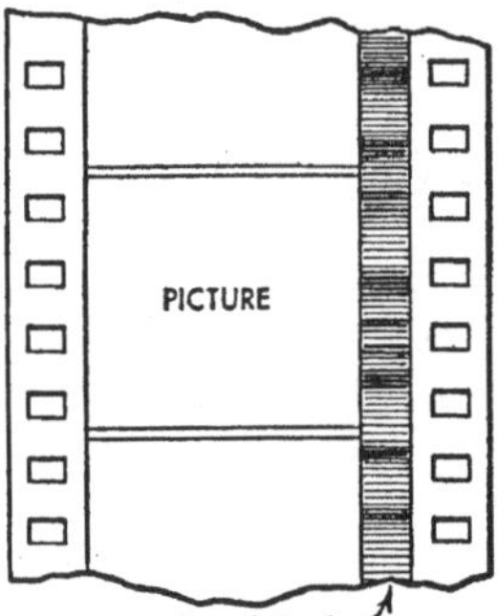


**Fig. 29.** Movietone sound-picture film

A positive print of this sound track is transferred to the motion-picture film as shown in Fig. 29, but with the sound associated with each picture located 14 1/2 inches ahead of the picture. This enables the film to run at constant speed for the sound but in jumps in front of the shutter and projecting lens. In the projection machine a constant source sends light through a narrow slit and the sound track to a phototube as shown in Fig. 30. The varying light received by the phototube reproduces photoelectrically the original current in the recording microphones. This current is amplified and sent backstage to a loud-speaker behind the screen.

**Fig. 30.** Reproduction of sound from the Movietone film

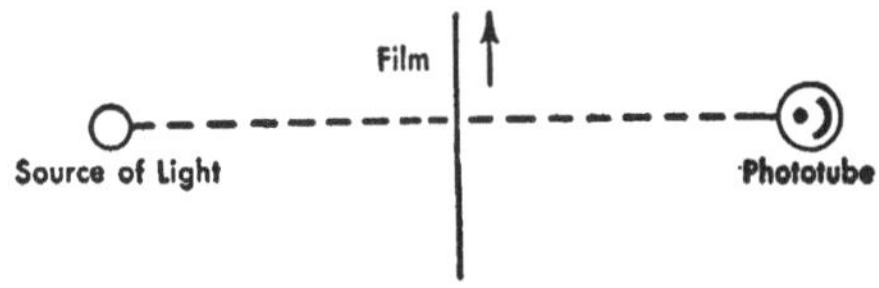


The sound track on the Photophone film is recorded by photographing the oscillations of a beam of light reflected from an oscillograph mirror. The oscillograph vibrator consists of a double strip of metal tape stretched over two bridges as shown in Fig. 31. This double strip is located between the poles of an electromagnet but in this case the direction of the magnetic flux is parallel to the plane of the double strip instead of perpendicular as in the light valve. When current is sent through the oscillograph vibrator its reaction with the magnetic flux will cause the double strip to twist through an angle proportional to the current.

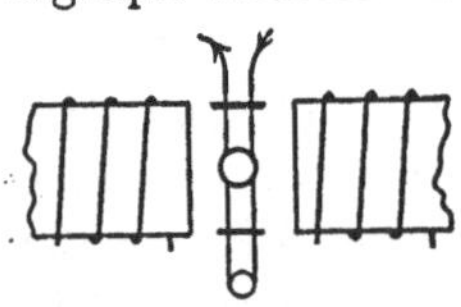

**Fig. 31.** Oscillograph vibrator

When the pictures are taken, light from a constant source is reflected from a mirror cemented on the double strip and through a narrow slit to the motion-picture film as shown in Fig. 32. Microphones pick up the sound and deliver the corresponding amplified current to the oscillograph vibrator. One side only of the sound

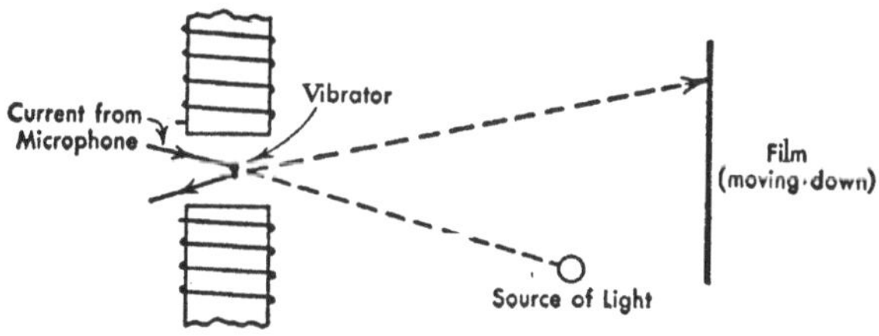


**Fig. 32.** Recording sound on the Photophone film

wave is photographed on the sound track as shown in Fig. 33. In the projection machine the variations in light transmitted through the sound track produce corresponding variations of current in a phototube as previously described. This current is amplified and delivered to a loud-speaker behind the screen. Either type of sound motion-picture film may thus be run in the same projection machine.

In some of the future sound-motion pictures it is probable that the sound will be reproduced from a separate film containing many sound tracks. The sound film will be run on a separate sprocket wheel mounted on the same shaft as the sprocket wheel for the picture film so that sound and picture will be accurately synchronized. With several sound tracks and associated loud-speakers the sound may be given a spatial effect. The audience may hear the sound or speech coming from the place where it is produced; either to the left or the right, or at the front or the back of the stage. A sound-motion picture of an orchestra may then be projected with greater realism because the sound will come from that part of the orchestra where the conductor points his baton. Many unusual results may be obtained by this type of reproduction. In one example already produced the voices of a choir come from the rear of the auditorium.

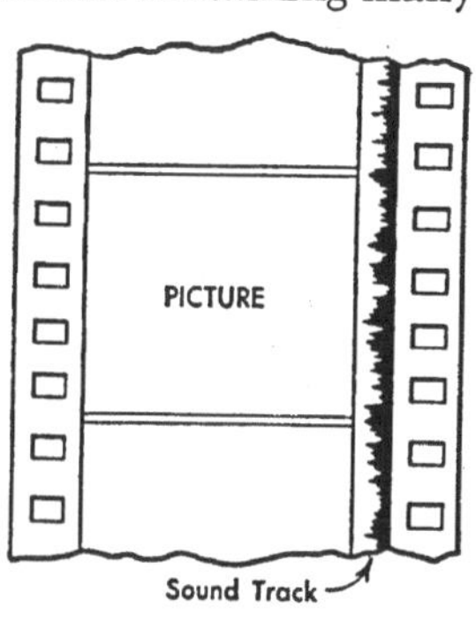


**Fig. 33.** Photophone sound-picture film

## Transmission of Still Pictures

The transmission and reception of photographs, maps, drawings, thumbprints, signatures, and hand- or type-written letters is usually called "facsimile" transmission and reception. When the reception

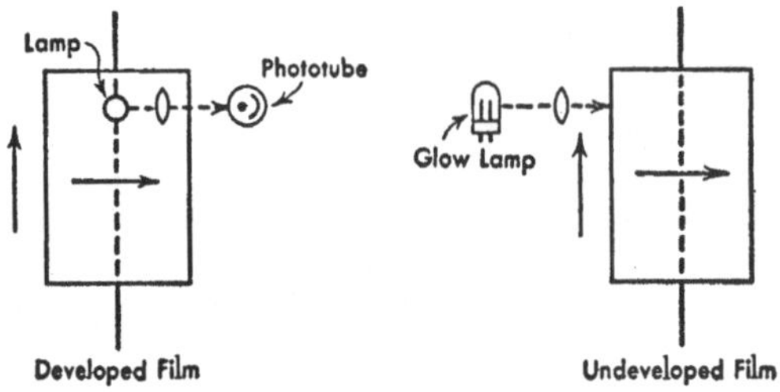


**Fig. 34.** Transmission of pictures by the **R**adio Corporation system

is obtained by a photographic method the system is often called "telephotography." In any case the actual transmission through space may be accomplished by wire or by radio.

In one of the photographic systems employed by the **R**adio Corporation of America the developed film to be transmitted is wrapped around a transparent cylinder which rotates and also moves longitudinally along its axis as shown in Fig. 34. A fixed and constant source of light on the axis of the cylinder sends a concentrated beam of light through the rotating film to a phototube outside of the cylinder. The intensity of light reaching the phototube varies (inversely) with the density of the photographic detail at each point on the film. This process is called "scanning."

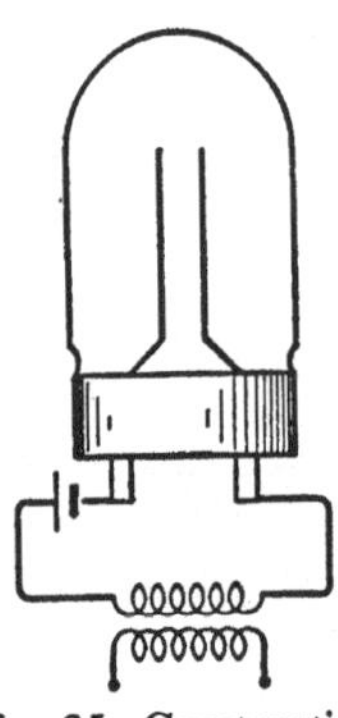

**Fig. 35.** Construction of a glow lamp

The current delivered by the phototube will vary in magnitude with the intensity of the light and may be amplified for transmission to the receiving apparatus. At the receiving end a glow lamp converts the varying current into light of corresponding intensity. The glow lamp consists of two parallel plates or disks enclosed in a glass tube, as shown in Fig. 35, and contains neon or helium gas at low pressure. If a direct voltage is applied between the plates, the lamp emits a brilliant light which varies widely in

*Radio Corporation of America*

A gas-type phototube used principally for sound reproduction.

*Finch Telecommunications, Inc.*

Picture sent from New York to San Francisco by wire and recorded photographically. The original is at the top and the reproduction at the bottom.

*Finch Telecommunications, Inc.*

A newspaper is delivered by wire or radio. In the recorder shown, the newspaper is printed photographically in a dark room but in another form the newspaper literally unrolls before your eyes in a lighted room.

*Finch Telecommunications, Inc.*

The photograph at the top is reproduced directly at any distance on electro-sensitive paper, as shown below. The texture of the reproduction resembles that of tapestry.

intensity with the magnitude of the voltage between the plates. A transformer, as shown in the figure, superposes a variable voltage on the direct voltage, the variable voltage being proportional to the current delivered from the phototube at the sending station. The light from the glow lamp is focused to a small spot on an undeveloped film (Fig. 34) wrapped on a cylinder which rotates and moves longitudinally (in a lightproof box) in synchronism with the sending cylinder. Both cylinders thus rotate and move along their axes together, and a negative of the original picture is produced on the film at the receiving station.

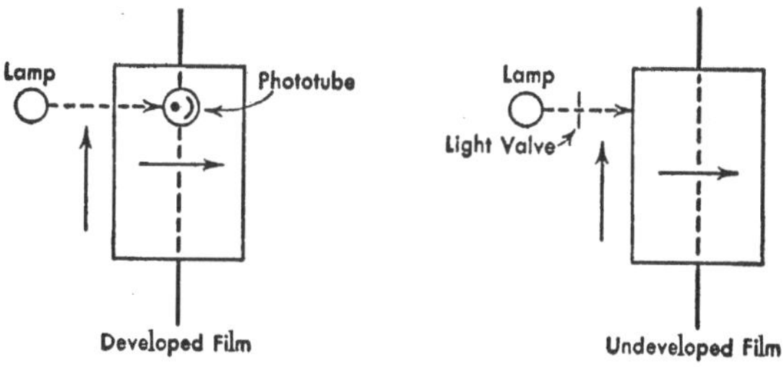


**Fig. 36.** Transmission of pictures by the Bell Telephone Laboratories system

In one of the systems developed by the Bell Telephone Laboratories the source of light is placed outside the sending cylinder, as shown in Fig. 36, and the phototube is placed on its axis. The amplified current from the phototube is transmitted to a light valve similar to that employed in the Movietone sound-recording apparatus described on page 49. The light from a constant source is focused through the light valve on the receiving film. The light transmitted by the light valve varies (inversely) with the density of the photographic detail on the sending film and produces a negative of the original picture on the receiving film. Pictures on opaque materials may also be transmitted by either system by reflecting concentrated light from successive light and dark spots on the picture to a phototube in which the current again varies (inversely) with the density of the spots.

While most systems of facsimile transmission employ a transmitter in which light is either transmitted through or reflected from a picture on a revolving drum to a phototube, many recording systems do not employ the photographic process. In the ink recorder, a very fine jet of ink is sprayed steadily from a nozzle, as shown in Fig. 37. A vibrating vane, actuated electromagnetically by the amplified current from the sending station, obstructs the ink jet for the white spots and allows the jet to strike the paper for the dark spots on the original picture. The paper or transparent film on the recording drum rotates and moves longitudinally at the same time.

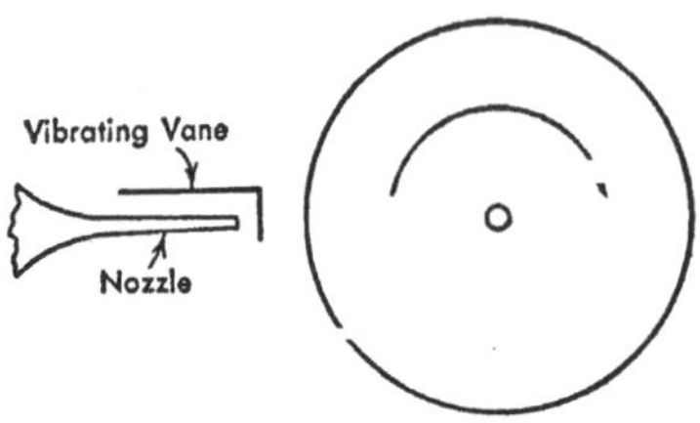


**Fig. 37.** Construction of an ink recorder

In electrolytic recording, the paper on the drum is impregnated with a chemical which turns dark when current passes through the paper. A metal stylus, which moves helically over all parts of the paper as it rotates, sends a pulse of current from the sending station through the paper for each dark spot on the original picture. In the carbon-paper recorder a piece of typewriter carbon paper is wrapped around the white paper fastened to the rotating drum. In this method, the stylus, actuated electromagnetically from the sending station, presses the carbon paper against the white paper for each dark spot on the original picture.

In most of the recording methods not employing the photographic process the reproduction is visible and subject to adjustment during the recording period. The finished picture is also immediately available without development. It will be noted that in the visible methods which make positive pictures, a black dot is printed when no pulse of current is received from the sending station. This inversion may be obtained in the electromagnetic relay at the receiving station or in the amplifiers at either end.

It may be expected some day that facsimile transmission will be used widely for the dissemination of news. That is being done currently by the various kinds of teletype which steadily print news, stock quotations, weather reports, and so forth. Facsimile transmission can do more. It can print in your home an entire newspaper containing news, pictures, cartoons comic strips, and advertisements. The paper will probably be tabloid in size and will be reproduced by one of the visible methods of reproduction. Such papers have already been transmitted and the general adoption of this service only awaits the construction of proper receiving apparatus at a reasonable cost.

## Television

Seeing at a distance involves a process in which we not only see any object placed before the transmitter but also see all the activity and animation upon which the transmitter is focused. At present there are two effective types of television transmitter: the Farnsworth "Dissector Tube" or "Orthiconoscope" and the Zworykin "Iconoscope." Both employ a photocathode containing a multiplicity or "mosaic" of photoemissive drops of cesium oxide spread evenly on a mica plate, as shown in Fig. 38. Since each drop of cesium oxide is insulated from the others the composite photocathode constitutes a myriad of condensers with one plate (the metal backing) in common.

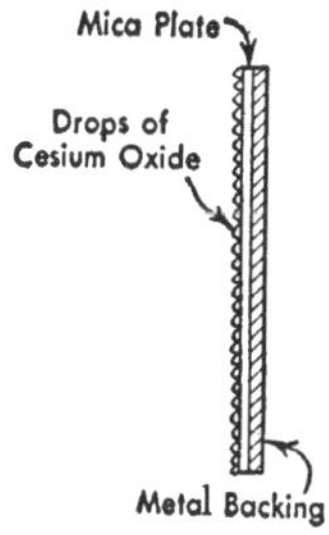


**Fig. 38.** Construction of a photocathode

In the Dissector Tube, shown in Fig. 39, the optical image focused upon the photocathode releases electrons from each element of the photocathode in direct proportion to the intensity of the light received by that element. These electrons are pulled to the left by the anode (a metal lining within the tube) and are maintained throughout the tube in the same relative position (as at the photocathode) by focusing coils surrounding the tube. Without

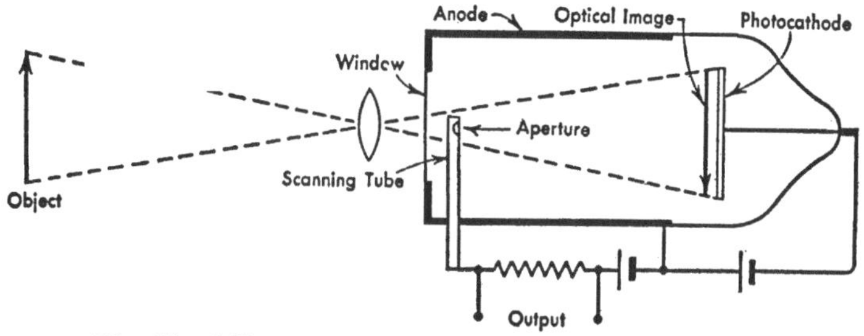


**Fig. 39.** A Farnsworth Dissector Tube or Orthiconoscope

further control the electrons emitted by the central drop of cesium oxide on the photocathode would pass through the aperture in the fixed scanning tube. The current output of the auxiliary anode within the scanning tube would then vary in proportion to the changes in light at a single central spot in the optical image.

Since moving electrons or an electric current may be moved in any direction by forces of electrostatic or electromagnetic origin, we may locate two sets of coils, perpendicular to each other, outside the tube and establish in them alternating currents of saw-tooth-wave form, which will sweep the entire electron image back and forth and up and down so that every point on the mosaic will shoot electrons successively through the aperture. In the usual scanning speed the electron image is resolved into 441 lines from top to bottom and the entire mosaic is covered in 1/30 second. The scanning speed is then 13,230 lines per second. Actually alternate lines are scanned, going down, in 1/60 second and the remaining lines are scanned, going up, in 1/60 second. The fixed scanning tube also contains an electron amplifier which greatly increases the magnitude of the output current.

In the Iconoscope the optical image is focused on a photocathode which also serves as the target or screen of a cathode-ray tube, as shown in Fig. 40. This type of cathode-ray tube is often called an

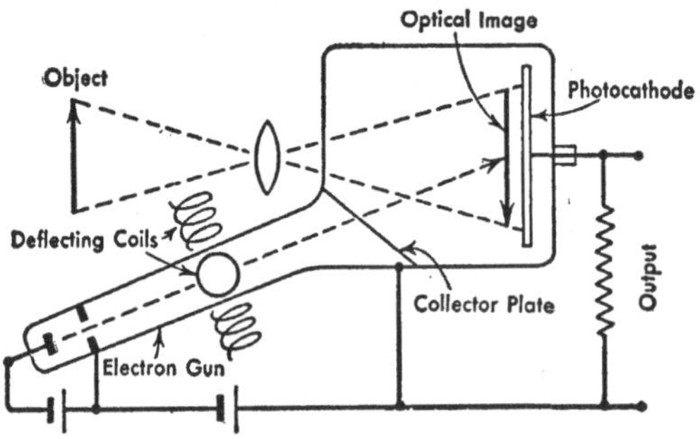


**Fig. 40.** A Zworykin Iconoscope

"electron gun." Sweep circuits located around the electron gun cause the cathode ray to scan every part of the electron image on the photocathode, 441 lines from top to bottom, 30 times per second. Although some electrons are added to each spot during the scanning process more electrons are lost due to the bombardment with the high-velocity electrons in the cathode ray.

In consequence the dark spots vary in voltage (above the collector plate) from − 1.5 to + 3 to − 1.5 volts during the scanning period; the + 3 volts for any spot is due to the loss of electrons (negative particles) each time the cathode ray sweeps by, and the − 1.5 volts is due to the absorption of excess electrons in the intermediate period from other parts of the mosaic. The illuminated spots, due to the additional loss of electrons photoelectrically, vary in voltage from + 1, say, to + 3 to + 1 volts during the scanning period. The current flowing in the resistor connected between the output terminals then varies inversely with the illumination of each spot and constitutes the television or "video" signal.

The television reproducer or receiver, called a "Kinescope" (Zworykin) and an "Oscillite". (Farnsworth) is substantially the same in either system. The pulsating signals from the transmitter are sent by wire or by radio to the negatively biased grid of a cathode-ray tube as shown in Fig. 41. The strength of the cathode

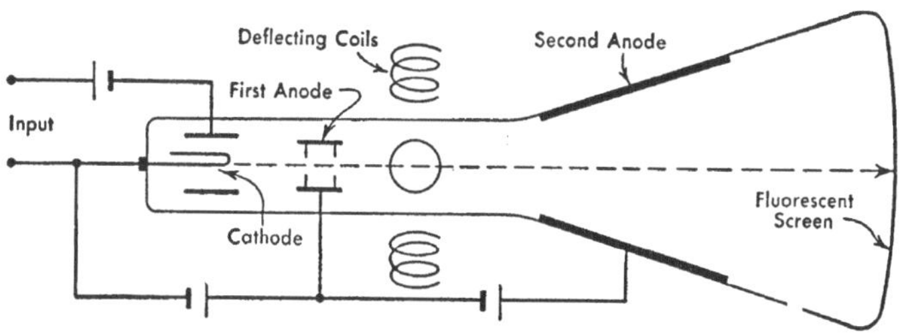


**Fig. 41.** A cathode-ray television receiver

ray is varied in direct proportion to the consecutive pulses of current supplied by the various elements of the photocathode at the transmitter. At the same time deflecting coils at the receiver sweep the cathode ray back and forth and up and down on the fluorescent screen in synchronism with the scanning speed at the transmitter.

Any scene in full animation may thus be transmitted and reproduced at a considerable distance from the source. Rotating disks transmitting the three primary colors in succession at the transmitter and the receiver will give reproduction in full color. Several transmitters focused at different angles on the original scene together with several cathode-ray tubes focused on the same screen of special construction will give reproduction in three dimensions with depth.

Television receivers may be produced with screens approximately 15 by 20 inches in dimensions and with sound accompaniment. Their general adoption will depend as usual upon the cost, which may be expected to be much greater than the cost of a receiving set for sound alone. Some degree of success has already been attained by reflecting the image from the fluorescent screen with mirrors to a larger white screen. The final objective, of course, is the enlargement of the image to the size of the screen in a motion-picture theater. Then we shall be able to sit in comfort and hear and watch the world in sound and motion at practically the very instant that the events take place.

CHAPTER V

# MODERN SOURCES OF LIGHT

## Radiation and Vision

The retina of the human eye responds only to an infinitesimal part of the radiation which it receives from a multitude of sources. For most people vision begins at a radiation frequency of $4.3 \times 10^{14}$ cycles per second (wave length of 7000 angstroms) and ends at $7.5 \times 10^{14}$ cycles per second (wave length of 4000 angstroms). The wave length of radiated energy is the distance traversed by the radiation in a vacuum in the time of one cycle. In centimeters, as previously stated, it equals $3 \times 10^{10}$ divided by the frequency. In angstroms it equals $3 \times 10^{18}$ divided by the frequency because an angstrom equals $10^{-8}$ centimeter. In this chapter all radiation curves are plotted against wave length in angstroms and the equivalent frequency in cycles per second, hereafter abbreviated "angs" and "cps," respectively.

The eye, moreover, does not respond equally over that limited region. From the relative visibility curve shown in Fig. 42 it will

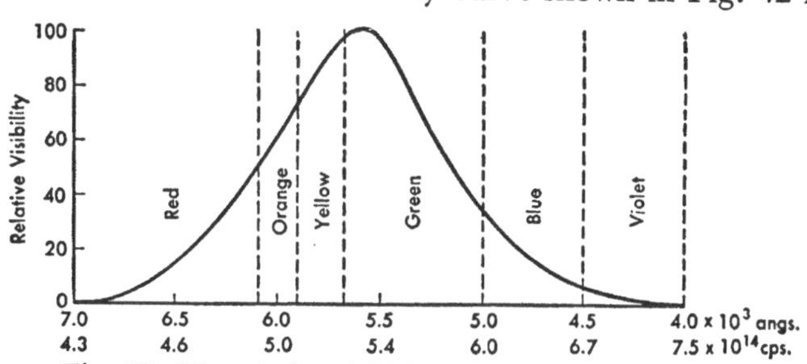


**Fig. 42.** The relative-visibility curve for the human eye

be seen that the average human being sees best at a frequency of $5.4 \times 10^{14}$ cps (5600 angstroms) or yellowish-green light. The color distribution shown is the usually accepted one although it is impossible to tell exactly when red turns into orange, orange into yellow, and so forth. No matter how intensely an object may be illuminated with low-frequency red light or high-frequency violet light it is not easily seen.

The radiation received from the sun at the zenith by a surface located at sea level is shown in Fig. 43. Since all parts of the visible spectrum are not equally effective and none of the infrared and ultra-violet radiation can be seen at all, it will be evident that only a small part of the radiation received from the sun is useful for vision. The proper frequency for vision is determined by three conditions.

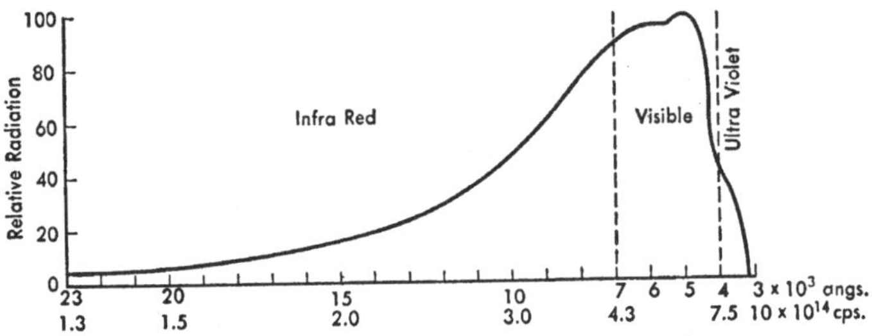


Fig. 43. Radiation received from the sun at the zenith by a surface located at sea level

For greatest visibility the light should be yellowish-green, a rather unpleasant color. If objects are to be seen in true color all frequencies of radiation in the visible spectrum must be received with the same relative intensity as that received from the sun. On the other hand, people look their best under an illumination which contains a slight excess of the red frequencies. This color is also desirable for the illumination in display cases of meat and many kinds of fruit.

Radiation at frequencies just outside of the visible spectrum is used advantageously for various purposes. Infrared frequencies ranging from 1.0 to $4.3 \times 10^{14}$ cps are effective in accelerating the drying of painted or lacquered surfaces, and for heat therapy or healing. Ultra-violet frequencies from 7.5 to $9.5 \times 10^{14}$ cps are used extensively in blue-printing and for the excitation of fluorescent paints. Frequencies from 9.5 to $10.5 \times 10^{14}$ cps are employed for the production of erythema or skin-tanning with the associated beneficial production of vitamin D. In the range between 10.5 and $13.5 \times 10^{14}$ cps, radiation has a germicidal effect and destroys bacteria in the air, on open wounds, on eating utensils, or on food.

*General Electric Company*

The largest and the smallest incandescent lamp. The 50,000-watt lamp is 3 feet high and has a bulb diameter of 20 inches. The 0.17 watt lamp (lower left) is 0.343 inch long and 0.079 inch in diameter. It is magnified about three times in the picture. If it were reduced to the same scale as the large lamp it would be only 0.017 inch in diameter.

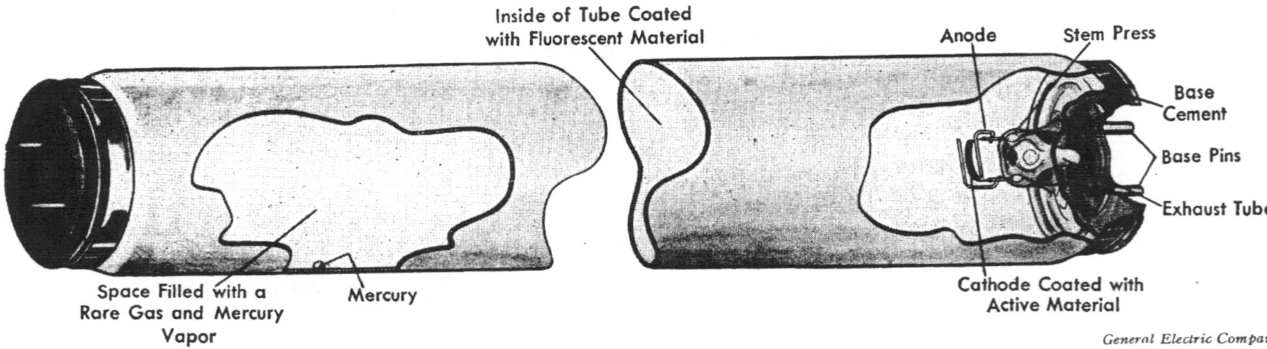


*General Electric Company*

Construction of a fluorescent lamp.

## Sources of Visible Radiation

The highest frequency current produced to date in an oscillator circuit (see page 42) is $3.6 \times 10^{12}$ cps, which is far below the visible spectrum. If a current of 4.3 to $7.5 \times 10^{14}$ cps frequency could be established in a coil of wire it would make an efficient source of light but no method is known for producing a current of that frequency. The usual artificial sources of the required frequency are "excited" atoms or possibly "unstable" molecules. In the normal or unexcited hydrogen atom the single electron rotates in the $n_1$ orbit as shown in Fig. 44. This is the least energy orbit or the lowest quantum level. If the electron is pulled by any outside influence into one of the outer orbits or quantum levels, the atom is said to be "excited" and the electron immediately falls back in a spiral path into the $n_1$ orbit but may stop momentarily in one of the intervening orbits.

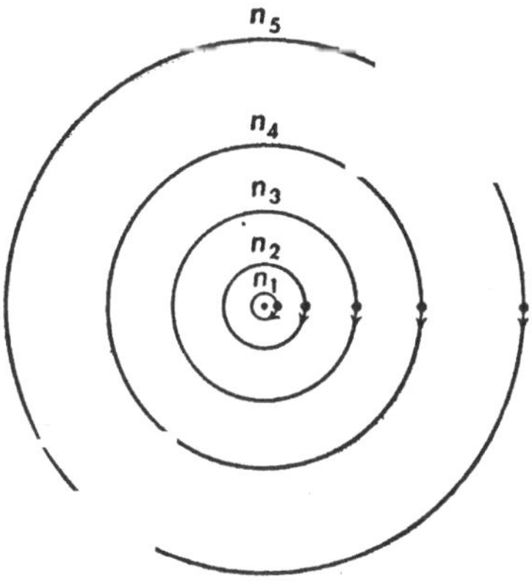


**Fig. 44.** The normal orbit ($n_1$), and some of the outer orbits ($n_2$, $n_3$, $n_4$, and $n_5$) of the electron in an excited hydrogen atom

If the electron falls without interruption from any outer orbit to the $n_1$ orbit, it will emit radiation at ultraviolet frequencies. When it falls only to the $n_2$ orbit the radiation will be largely in the visible frequencies, but when it falls to the $n_3$ orbit or any other outer orbit the radiation will be in the infrared. Radiation is produced similarly in any atom under excitation when electrons fall from outer to inner orbits. Although not clearly understood, it is believed that a molecule emits radiation when its constituent atoms are changed from their normal relative positions by external influences. Although there are many other causes, excitation of atoms in solids or liquids is usually caused by the absorption of heat and in gases by the momentary ionization of some of the constituent molecules. In the following discussion it will be shown that excitation in a gas by ionization is more efficient than excitation in a solid or liquid by heating.

## Production of Light by Incandescent Solids

The radiation from an incandescent solid or liquid is principally in the infrared spectrum and is invisible. Production of a small part of the radiation in the visible spectrum is accomplished only by raising the temperature of the radiating substance to the highest possible value. This requirement precludes the use of incandescent liquids because any transparent container would melt at such a high temperature. The filament of an incandescent lamp must then have a very high melting point and also possess sufficient mechanical strength to maintain its form when operated at a high temperature. Since carbon has the highest melting point among the elements — it actually changes from a solid to a vapor at 3600 degrees Centigrade — it might appear to be best suited for the purpose. A carbon filament, however, begins to evaporate at a comparatively low temperature and fills the enclosing bulb with carbon vapor which condenses on the inside of the bulb and shuts off much of the light.

Tungsten, on the other hand, has a melting point (3380 degrees Centigrade) close to that of carbon and has a much lower vapor pressure so that it may be operated at a higher temperature than carbon with less blackening of the bulb. Other requirements of a desirable material for a lamp filament are high resistivity, high positive temperature coefficient of resistance, good ductility, and moderate cost. If the resistivity is high, the filament may be comparatively thick and short and therefore strong. A high positive temperature coefficient of resistance indicates an increase in resistance with the temperature so that a moderate increase in lamp voltage above normal will not produce a destructive increase in the current because the resistance will also rise with the voltage. Tungsten satisfies all of the requirements except that its resistivity is so low that the filament of a small lamp must have a diameter of not more than a few thousandths of an inch. The filament is nevertheless rugged, because tungsten has a very high tensile strength; about four times that of steel when cold.

The operating temperature of a tungsten filament varies from 2130 degrees Centigrade (6-watt lamp) to 3060 degrees Centigrade (10-kilowatt lamp). The usual 60-watt lamp has a temperature of 2500 degrees Centigrade, which is sufficient to melt as-at 115 volts on either alternating or direct current. The direct-current connections are shown in Fig. 47 *a* and the alternating current in Fig. 47 *b*. Since the tube acts as a rectifier (as explained in the next chapter), the alternating-current tube must have two anodes connected to either side of the line as shown in the figure. In either case the tube must be connected in series with "ballast" (a resistor for direct current, and a resistor, inductor, or capacitor for alternating current) to stabilize the current in the tube. In gaseous conduction an increase of current, if uncontrolled, would cause increased ionization and a consequent further increase of current which in a short time might attain sufficient magnitude to burn out the tube. An increase of current in the ballast, however, reduces the voltage across the tube and maintains the current at a constant value.

Gaseous conduction tubes with their characteristic colors may be bent in any form for decorative or advertising purposes. The mercury-vapor tubes with liquid mercury cathodes have been employed extensively for industrial and commercial lighting in plants where discrimination of color is not important.

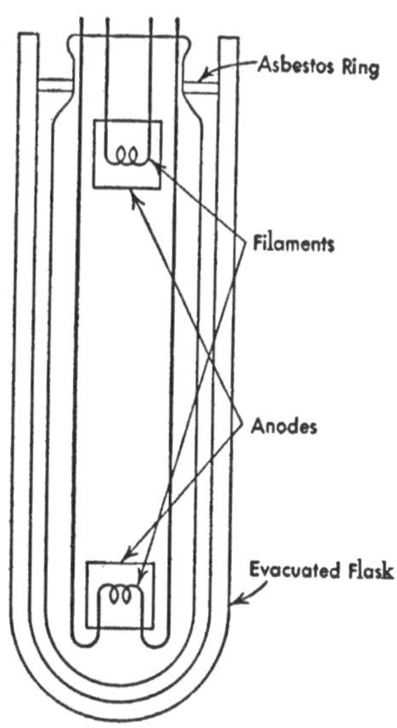

**Fig. 48.** A sodium-vapor lamp

If the cold electrodes of gaseous-conduction tubes are replaced by heated filaments, the applied voltage may be reduced to the usual house-circuit values or even less. In the sodium-vapor lamp shown in Fig. 48 a short tube containing neon gas and a small amount of liquid sodium (when cold) is enclosed by a second double-walled glass tube in which the intervening space is highly evacuated to prevent the loss of heat. The filament electrodes at either end of the inner tube are first connected in series across a transformer until they are heated; an automatic switch then breaks the connection between them (as explained later in connection with the fluorescent lamp) and impresses sufficient voltage between the heated electrodes and their associated anodes to ionize and excite the neon gas. After a few minutes the inner tube attains a temperature sufficient to evaporate the liquid sodium into sodium vapor, which by ionization and excitation produces the characteristic yellow light. While the color is not suitable for general purposes this lamp has been applied extensively to highway lighting.

A considerable portion of the power input to most types of gaseous-conduction lamps is radiated internally in the ultraviolet spectrum and is absorbed by the glass tube. Various "fluorescent" materials are excited by ultraviolet radiation so that electrons within their atomic structure when returning to lower energy levels will radiate energy at visible frequencies. Calcium tungstate radiates blue light; zinc silicate, green light; cadmium borate, red light, and so forth. If the inside of a mercury-vapor tube is coated with one of these fluorescent materials, called "phosphors," a considerable portion of the otherwise wasted ultraviolet radiation will be converted into visible radiation and transmitted through the glass tube. By choosing various mixtures of the phosphors and using different colors of glass tubing practically any color of light may be obtained in the modern "fluorescent" lamp.

In the green lamp the efficiency is 18.5 per cent, which is about 2.5 times that of the usual gas-filled incandescent lamp. Phosphors may be employed which will reproduce any color of daylight (see Fig. 45) so that fabrics may be matched and wool or leather may be sorted under artificial illumination. For house, office, and store illumination a white light with a slight rose tint has proved most popular.

When a fluorescent lamp is turned on, current is established in the circuit shown in Fig. 49, which includes the two filament elec-

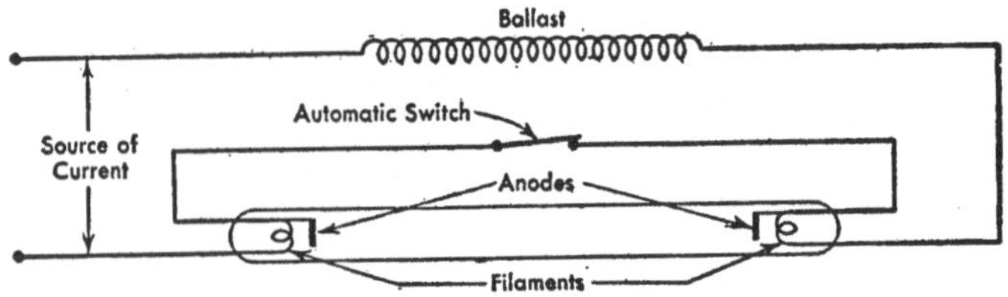


**Fig. 49.** A fluorescent lamp with automatic switch and ballast

trodes, two anodes, a closed automatic switch, and the ballast. When the electrodes are heated sufficiently the automatic switch opens and the consequent increase in voltage between the electrodes starts gaseous conduction within the tube between the anodes. While the source of current may be either direct or alternating the fluorescent lamp operates more efficiently on alternating current because inductor or capacitor ballast absorbs less power than resistor ballast.

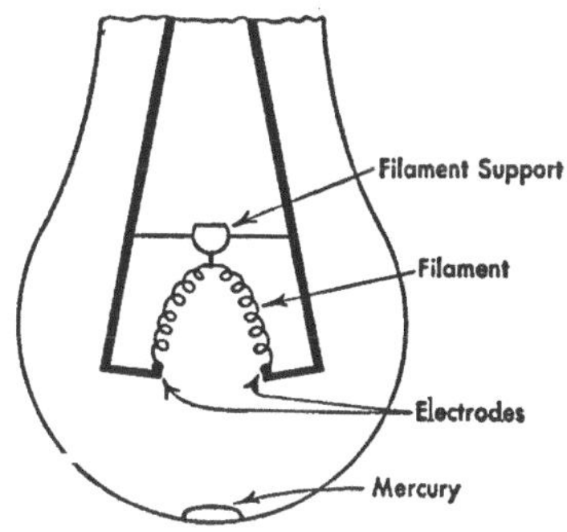

Fig. 50. A Sunlight lamp

Ultraviolet radiation for skin-tanning in the 9.5 to 10.5 $\times 10^{14}$ cps region may be produced by the Sunlight lamp, shown in Fig. 50, which contains a tungsten filament, a shunted gap, and some mercury, all enclosed by a bulb made of special glass which transmits the proper frequency and strength of ultraviolet radiation for beneficial skin-tanning. When the lamp is turned on the heated filament evaporates some of the mercury, and gaseous conduction is established between the ends of the filament. In one type of Sunlight lamp a mild sunburn will be produced on untanned skin at a distance of 30 inches in about 7 minutes.

Flashes of light of extremely short duration may be obtained by gaseous-conduction "flash" lamps of various types. In one form shown in Fig. 51, the flash lamp consists of a hollow glass tube,

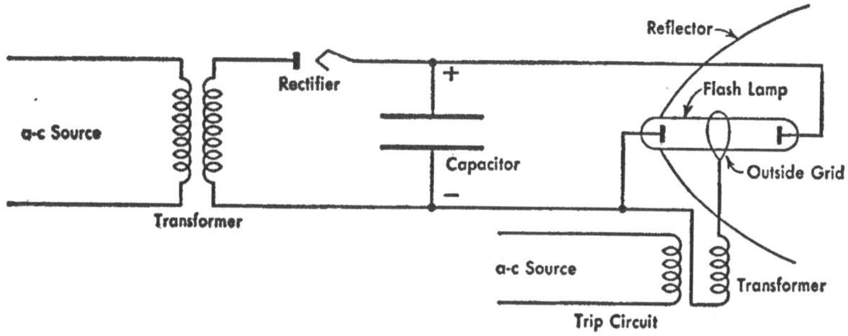

Fig. 51. A flash lamp with capacitor storage and a trip circuit

2 inches long with an outside diameter of 1/4 inch and an inside diameter of 1/6 inch filled with argon gas and a small amount of hydrogen at a pressure of about 1/2 atmosphere. A capacitor, connected between the cold terminals of the lamp, is charged from

an alternating-current source by a transformer and a diode rectifier to a voltage of 7000 volts. This voltage is slightly less than the amount required to discharge the capacitor through the tube.

When the trip circuit is closed, a small transformer impresses 15,000 volts on the outside grid, which is a single turn of wire wound around the flash lamp. The capacitor then discharges through the flash lamp in two-millionths of a second with an initial current of 1000 amperes. The initial power supplied to the lamp is then 7000 volts times 1000 amperes or 7,000,000 watts (7000 kilowatts). The lamp thus produces a brilliant flash of light in two-millionths of a second which is equivalent at least to the combined light given by 70,000 100-watt incandescent lamps.

Flash lamps are unsurpassed for making photographic portraits because the subject need not remain still and the flash is too quick to affect the eyes. Instantaneous photographs may also be made of high-speed objects and machinery. Industrial studies may thus be made of changes in form or position of the various parts of a machine while in action. In the stroboscopic application of the flash lamp, a machine may be illuminated by a rapid succession of accurately timed flashes so that any part may appear to be at rest or to move with a speed slow enough to be inspected. In this manner adjustments may be made under observation so that a machine may be operated at the highest efficiency with a minimum vibration and wear.

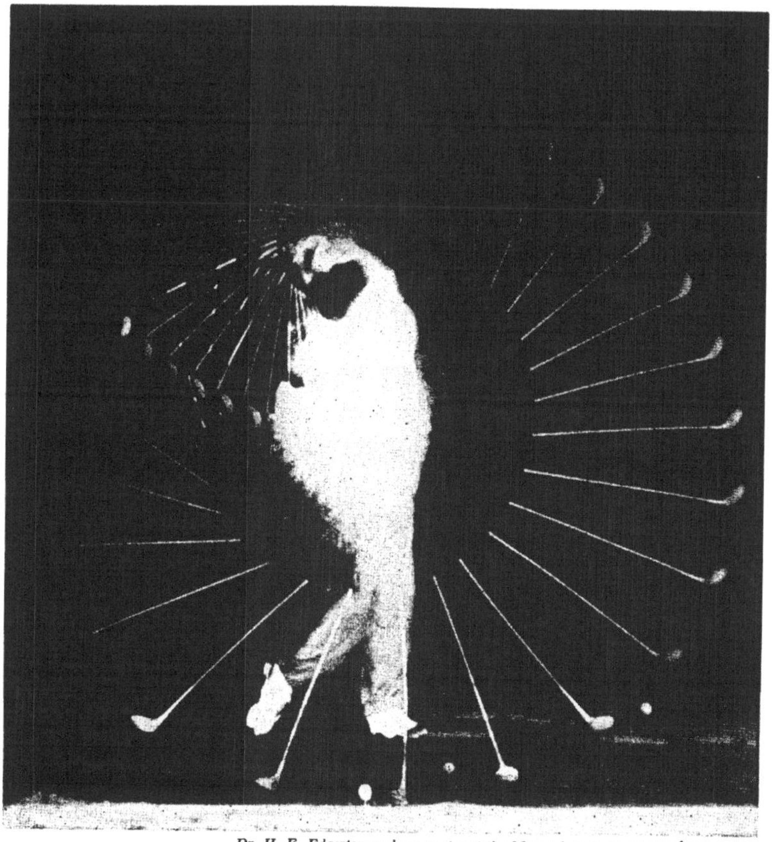

*Dr. H. E. Edgerton and associates at the Massachusetts Institute of Technology*

A multiple-exposure picture of one swing of a golf club taken with a flash lamp. The lamp flashes every 0.01 second and shows the golf club in fifty equally timed positions. It will be noted that the ball is traveling twice as fast as the head of the golf club at the bottom of the swing.

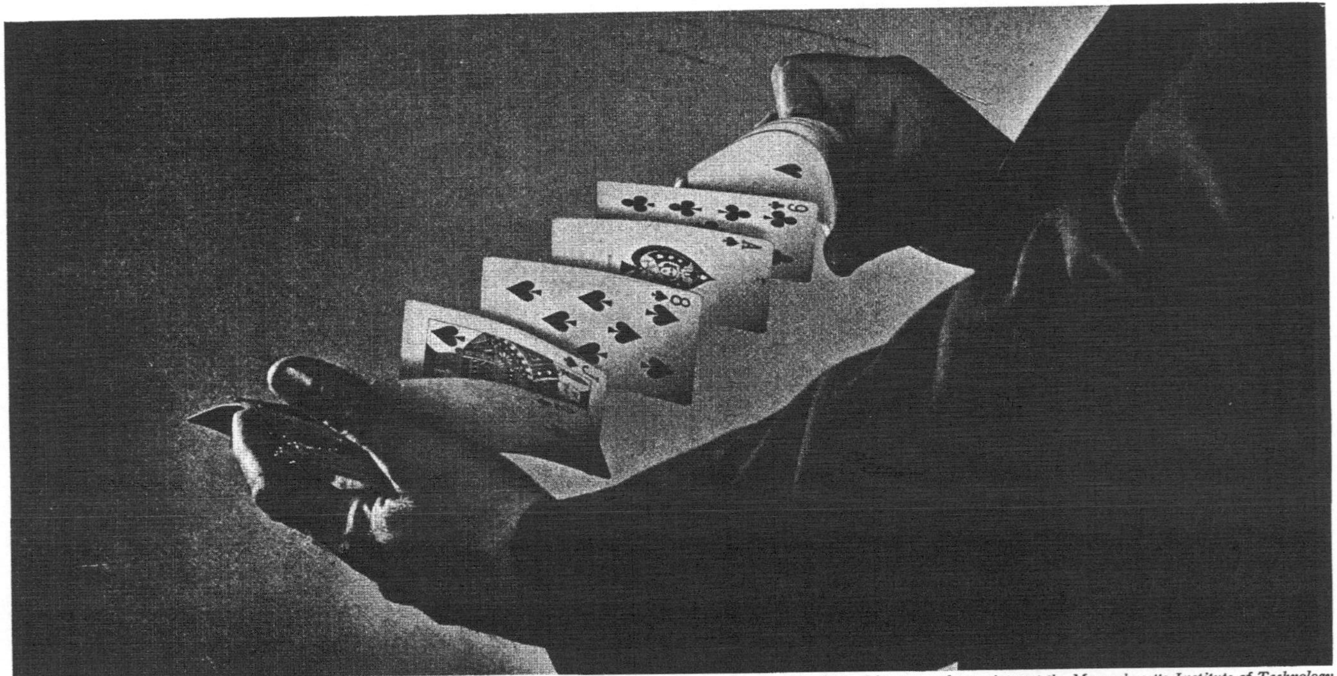

*Dr. H. E. Edgerton and associates at the Massachusetts Institute of Technology*

A magician "springs" a pack of cards from one hand to the other. The flash-lamp exposure of 1/100,000 second makes the cards appear to be standing still.

*Dr. H. E. Edgerton and associates at the Massachusetts Institute of Technology*

Although beating 60 times per second while hovering, the motion of a hummingbird's wings is stopped by the flash lamp. Flash-lamp pictures have shown that the hummingbird may fly backward as well as forward and steps up the beat to 70 times per second when flying. The sharp point at the end of the bill is the tongue. This type of hummingbird weighs 0.1 ounce.

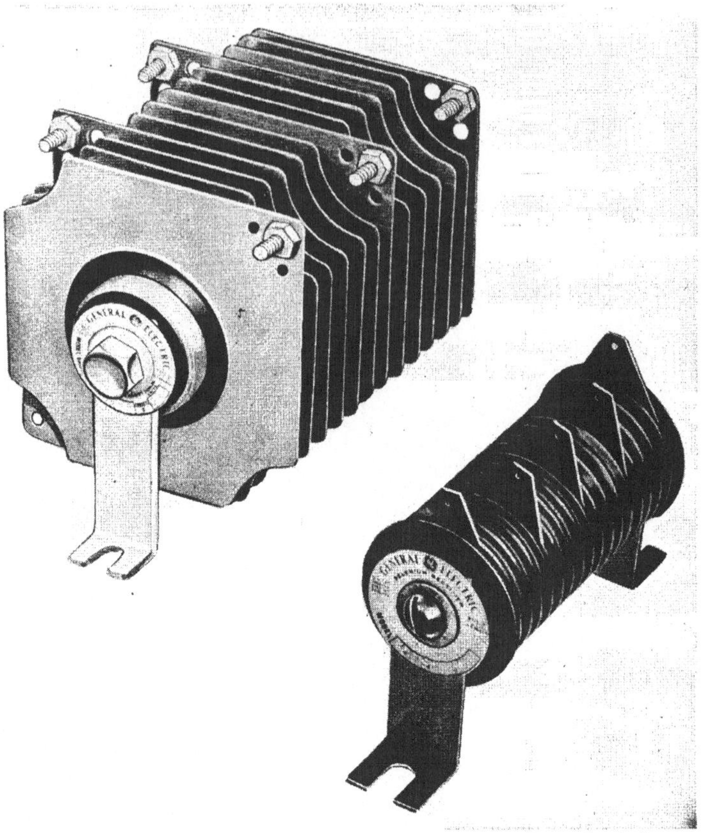
*General Electric Company*

Copper oxide rectifier, upper left, and selenium rectifier, lower right.

CHAPTER VI

# MORE POWER TO THE ELECTRON

## From Watts to Kilowatts

Although most of the applications of electronics described in the previous chapters have required relatively small amounts of power, the operation of many electronic devices is dependent upon the absorption or delivery of power in large quantities. The power input to many broadcasting antennas is as much as 50 kilowatts (12.5 amperes at 4000 volts, for example). Radiotelephone stations transmitting overseas often require as much as 500 kilowatts for satisfactory reception under adverse conditions. Since electric power may be generated and transmitted more economically by alternating current than by direct current under present limitations, heavy-duty electronic devices have also been developed to convert alternating current to direct current at the load end when the former current will not serve as well or not at all.

## Applications of Direct Current

In any electrolytic process, such as electroplating, or charging of storage batteries, direct current is indispensable because alternating current would not function. Direct current is superior for arc lighting and for many forms of metallic-arc welding. When industrial machines require accurate speed adjustment, a direct-current motor drive is usually preferred. Almost all electric railways are operated by direct-current motors because they are well adapted to the task of accelerating large masses to a high speed in a short time. Lifting magnets and magnetic relays of all kinds are usually built for direct-current consumption because they are smaller and more efficient. Except for bell ringing and multiplex transmission by wire, a telephone system employs direct current throughout.

## Rectifiers of Moderate Currents

The high-vacuum diode described on page 29 may be employed as a rectifier of currents ranging up to 100 amperes. At the higher values of current the diode must be water-cooled and for small

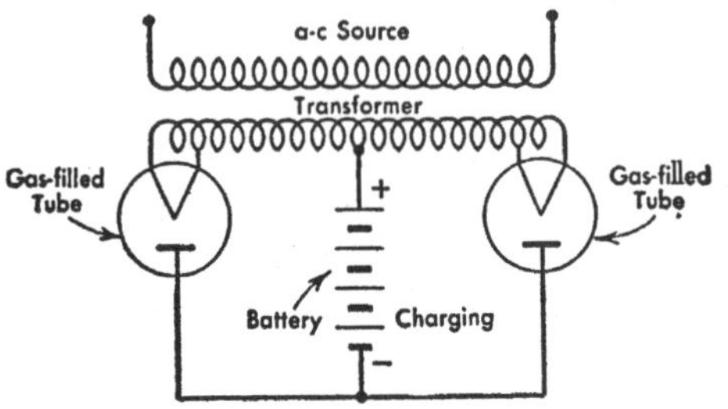


**Fig. 52.** A full-wave, gas-filled tube rectifier

values is best adapted to the rectification of currents with a voltage as high as 150,000 volts. A diode containing mercury vapor or argon gas at a pressure ranging from 0.1 to full atmospheric pressure will rectify currents up to 15 amperes at 115 to 230 volts. The increased current at low voltage is due to the ionization of the vapor or gas. Full-wave rectification is obtained by the connection shown in Fig. 52. This is the familiar Tungar or Rectigon rectifier employed extensively in garages for charging storage batteries.

## Grid-controlled Rectifiers

A gas-filled triode, called a "Thyratron," contains a heated cathode which is surrounded by a perforated grid as shown in Fig. 53. As explained on page 31, no current will flow through the tube (anode to cathode) if thermionic emission is suppressed by a negative grid voltage of sufficient magnitude. Rectification is started by reducing the negative grid voltage so that the tube functions both as a rectifier and a switch without moving contacts. A direct-current shunt motor may be supplied through four Thyratrons (two for the armature and two for the field) from an alternating-current source and may drive an automatic machine at closely controlled speeds with frequent starts and stops. After gaseous conduction is started in any

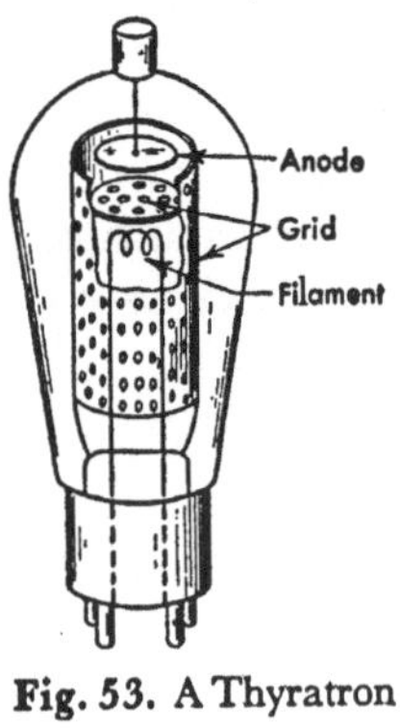


**Fig. 53.** A Thyratron

Thyratron by reducing the negative grid voltage it may not be stopped by restoring the original negative grid voltage. Gaseous conduction ceases only when the plate voltage decreases to a magnitude less than the critical value required to maintain conduction. Thyratrons will rectify a current as high as 100 amperes at 115 volts and up to 12.5 amperes at 10,000 volts.

In another application of the Thyratron a direct current may be converted to an alternating current of any desired frequency in

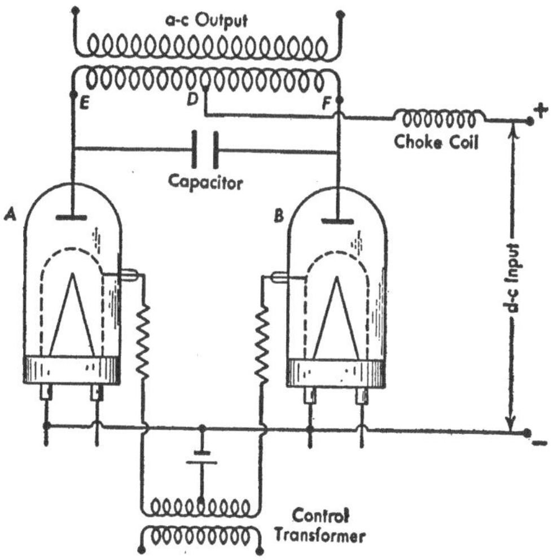


**Fig. 54.** One form of inverter which converts direct to alternating current

various forms of apparatus called "inverters." In one type shown in Fig. 54, a control transformer, connected to an auxiliary alternating-current source of the desired frequency, will start the Thyratrons (*A* and *B*) alternately so that the direct-current source will send current consecutively from *D* to *E* and from *D* to *F*. An alternating current is thus induced in the secondary winding of the power transformer with the same frequency as that of the control transformer and with any desired voltage.

## The Glass Mercury Arc Rectifier

A highly evacuated glass chamber (Fig. 55) contains two recessed anodes, $A_1$ and $A_2$, a cathode, *C*, and an auxiliary starting anode, $A'$. The anodes $A_1$ and $A_2$ are made of graphite or iron and the cathode consists of a pool of mercury. The starting anode, $A'$, also contains a smaller pool of mercury. The anodes $A_1$ and $A_2$ are connected to the opposite lines of the alternating-current source. The starting anode, $A'$, is connected through a resistor to one of the lines. The device in which direct current is to be established is connected between the cathode and the midpoint, $D$, of an inductor connected between the lines.

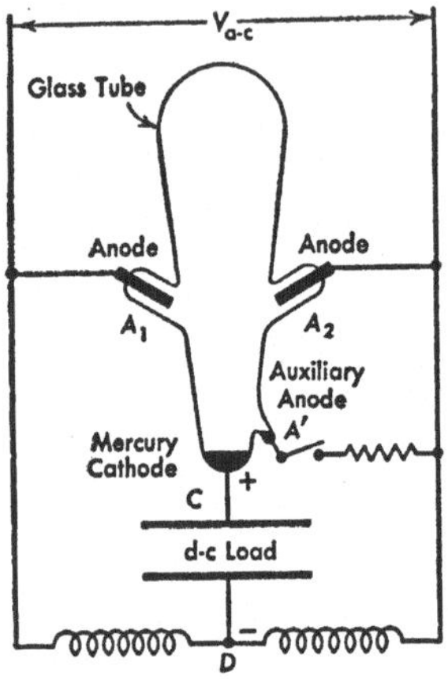


**Fig. 55.** A glass mercury arc rectifier

If the tube is tipped until mercury flows from $A'$ to $C$, an alternating current flows between $A'$ and $C$ and establishes an arc which rises to the anodes $A_1$ and $A_2$. Electrons will then flow alternately from $C$ to $A_1$, and from $C$ to $A_2$. Since the direction of each current is opposite to that of the electrons, current will flow from $A_1$ to $C$, and then from $A_2$ to $C$. The current flowing in any device connected between $C$ and $D$ will then be a direct current with the positive terminal at $C$ and the negative terminal at $D$. This type of rectifier will deliver a direct current as high as 50 amperes at 115 volts and at voltages as high as 15,000 volts with smaller currents.

## The Steel-tank Mercury Arc Rectifier

Much heavier currents may be rectified in the water-cooled steel-tank rectifier shown in Fig. 56. Vacuum pumps under continuous operation keep the pressure within the rectifier at the proper low value. Another pump circulates water through the jacket and an outside radiator. Emission of electrons from the mercury cathode is started by a high-voltage spark produced by an induction coil. The action within the tank is otherwise the same as that described in the glass type. Steel-tank rectifiers are used principally for the conversion of alternating-current to direct-current power for electrolytic plants and electric railways. They possess a current capacity as high as 5000 amperes at 625 volts and for short periods (1 minute) will rectify as much as 15,000 amperes. High-capacity rectifiers usually have six anodes connected to a 6-phase, 60-cps, alternating-current source in which each anode is supplied with current 1/360 second after the preceding anode and thereby produces a steadier direct current.

In the Ignitron mercury arc rectifier each anode is located in a separate tank which also contains an ignitor, a rod of high-resistance refractory material with a conical tip slightly immersed in the mercury pool and connected through an upper sealed electrode to the anode of a Thyratron. When the anode in any tank starts to become positive the Thyratron immediately increases the voltage between the ignitor and the mercury pool. The surrounding vapor is then ionized and an arc is established between the anode and the mercury pool. With such accurate timing the rectifier operates more efficiently with reduced interruptions. In another type, called an "Excitron" rectifier, each tank contains a continuous pilot arc which similarly produces accurate timing.

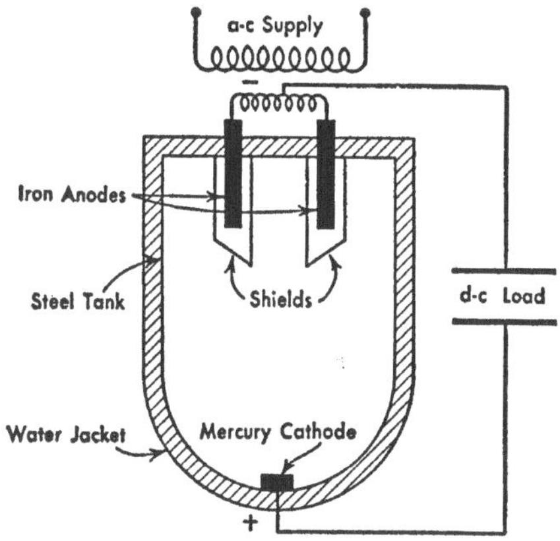


**Fig. 56.** A steel-tank mercury arc rectifier

## Rectox and Selenium Rectifiers

In the Rectox, or copper oxide rectifier, a circular disk or washer of copper is oxidized on one side at a temperature near the melting point, forming a thin layer of cuprous oxide. An adherent plating (the counter electrode) of lead, tin, and bismuth, or cadmium, tin, and thallium is then sprayed in hot, liquid form on the layer of cuprous oxide as shown in Fig. 57. When an alternating voltage is impressed between the counter electrode and the copper backing, electrons flow freely from the copper to the counter electrode but only to a minor degree in the other direction. The current thus flows principally from the counter electrode to the copper backing and the device constitutes a simple form of rectifier.

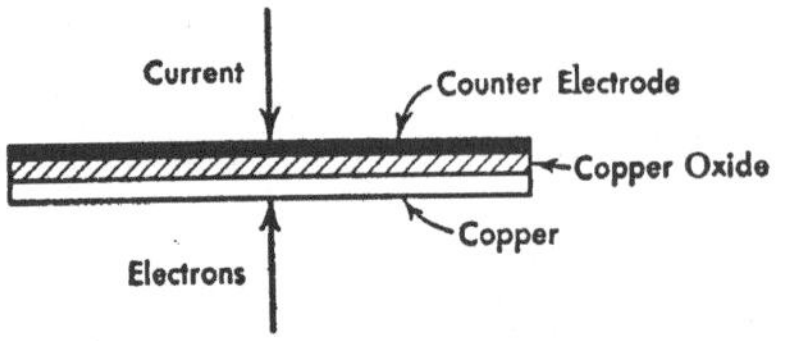


**Fig. 57.** A copper oxide rectifier

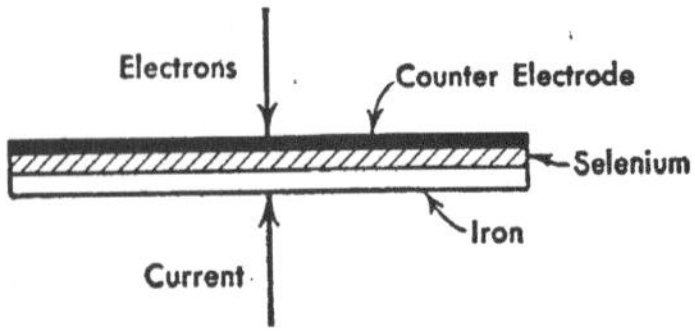


**Fig. 58.** A Selenium rectifier

In the Selenium rectifier, shown in Fig. 58, powdered selenium is compressed at a moderate temperature on one side of an iron disk and is then sprayed with an alloy as described above, which forms the counter electrode. In this type the electrons flow freely from the counter electrode to the iron disk but only slightly in the other direction. This device also serves as a rectifier but with current flowing from the iron disk to the counter electrode which is opposite to the direction of the current in the Rectox rectifier. Either rectifier has a capacity of about 0.3 ampere per square inch and may be connected in large numbers in parallel for rectification of large currents at low voltage or may be connected in series for rectification of small currents at high voltage.

## Photovoltaic Cells

When constructed with translucent conducting counter electrodes, or metallic circumferential rings, which leave the cuprous oxide or selenium layers exposed to radiation, the Rectox and Selenium rectifiers may be employed as direct converters of radiation to electric power. With the sun at the zenith on a clear day the radiation received by an area of a square foot at sea level is 75 watts. The amount of energy received by the earth on half of its surface per hour is the equivalent of that obtained by the perfect combustion of 21 billion tons of coal. The possibility of utilizing this major source of energy and storing it from the output of photovoltaic cells is a matter of great importance.

Unfortunately the cells available at the present time have an efficiency of only 0.5 per cent so that very little of the energy may be recovered. There is no reason, however, why future research may not develop new photovoltaic materials of sufficient efficiency to make their application economically desirable. Then we may expect the roof of every house to face southward and be shingled with photoelectric cells which will send current during sunlight hours to an electrically heated boiler in the basement. From this boiler may be drawn all or part of the energy required to heat the house, produce refrigeration, and, through a small turbo-generator, supply electricity for lighting and household appliances. A properly insulated boiler would be more efficient than a storage battery for the purpose. The principal application at the moment is a foot-candle meter made to measure the illumination on any surface in or out of doors. It consists of a photovoltaic cell connected in series with a microammeter. The scale is calibrated so that the pointer indicates the illumination directly in foot-candles.

CHAPTER VII

# DIVERSE APPLICATIONS OF ELECTRONICS

## The Electron Microscope

At the least focusing distance of 10 inches the average human eye cannot separate or "resolve" two dots or parallel lines located less than about $5 \times 10^{-3}$ inch apart. There are several reasons for this limitation. In the first place the image of a point formed by any lens is not a point but a disk surrounded by rings. If the two

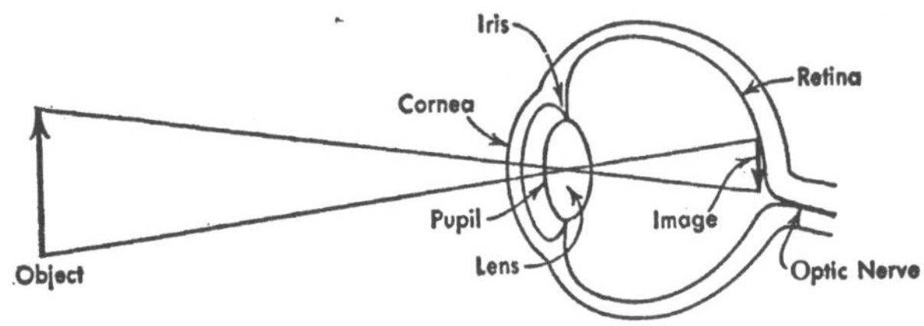


**Fig. 59.** The function of the human eye

points are very close together, the rings will overlap and blur the image. The two points will not appear to be separated unless the disk in one image falls no closer to the other disk than the inner ring of the second image. Since this distance is a direct function of the wave length of the light, resolution by any system of lenses improves as the wave length is shortened. Another limitation (see Fig. 59) is imposed by the coarseness of the grain structure of the retina within the eye. No sense of separation of points will be conveyed by the optic nerve to the brain unless their images fall upon separate nerve terminals. Small dots or fine lines separated by less than the critical distance will appear as a uniform gray surface.

If a double-convex lens (a reading glass) is placed between the eye and the object (see Fig. 60) the eye sees a greatly enlarged image of the object so that the object may appear to be magnified as much as 100 times. Magnification alone will not reveal the fine structure of the object unless the various points on the retinal image are sepa-

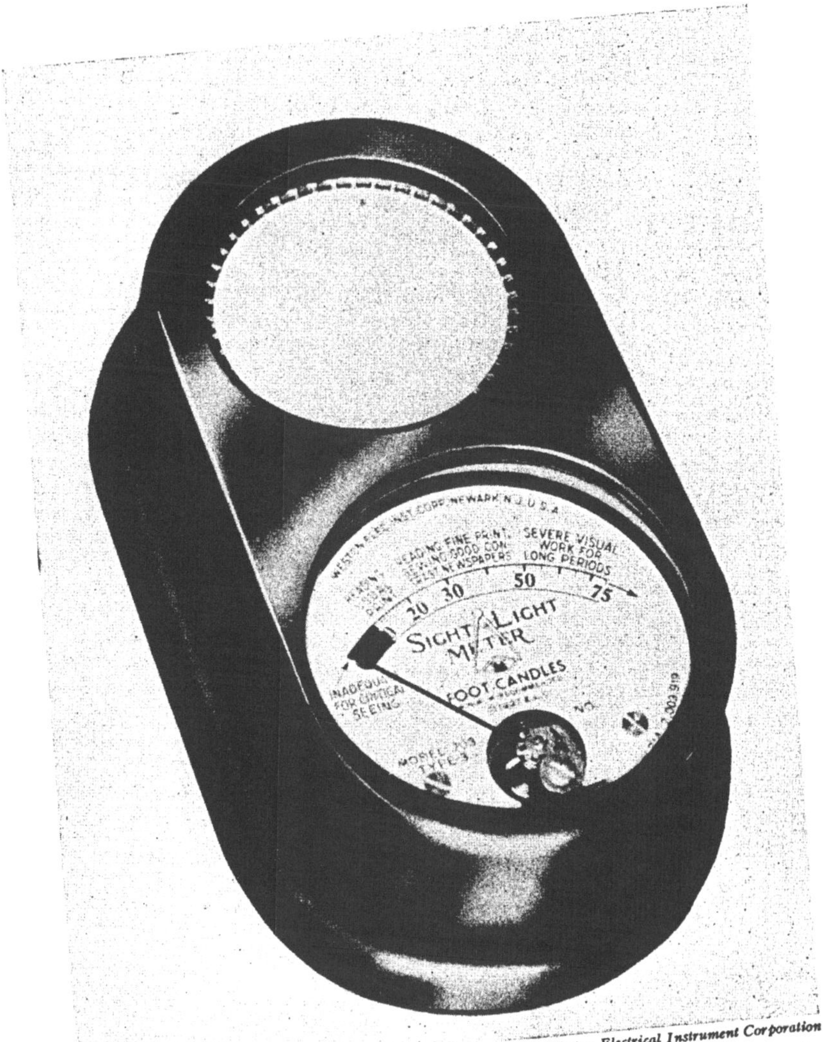


*Weston Electrical Instrument Corporation*

A sensitive ammeter connected to a photovoltaic cell indicates directly the illumination on the surface of the cell. The source of illumination supplies the power to operate the instrument.

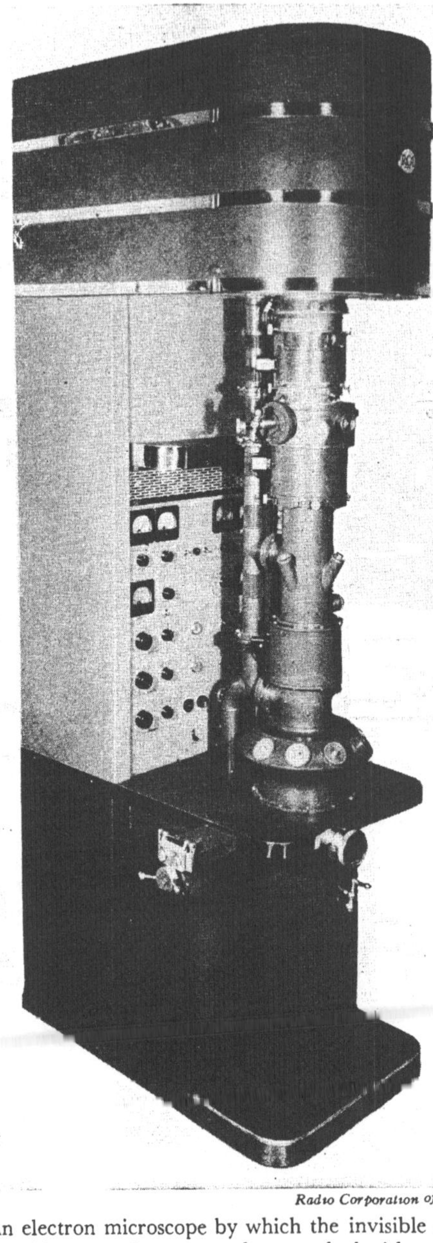

*Radio Corporation of America*

An electron microscope by which the invisible details of an object may be seen or photographed with a magnification of 100,000 diameters. After removing the circular caps located around the base of the instrument several people may look at the image on the fluorescent screen at the same time.

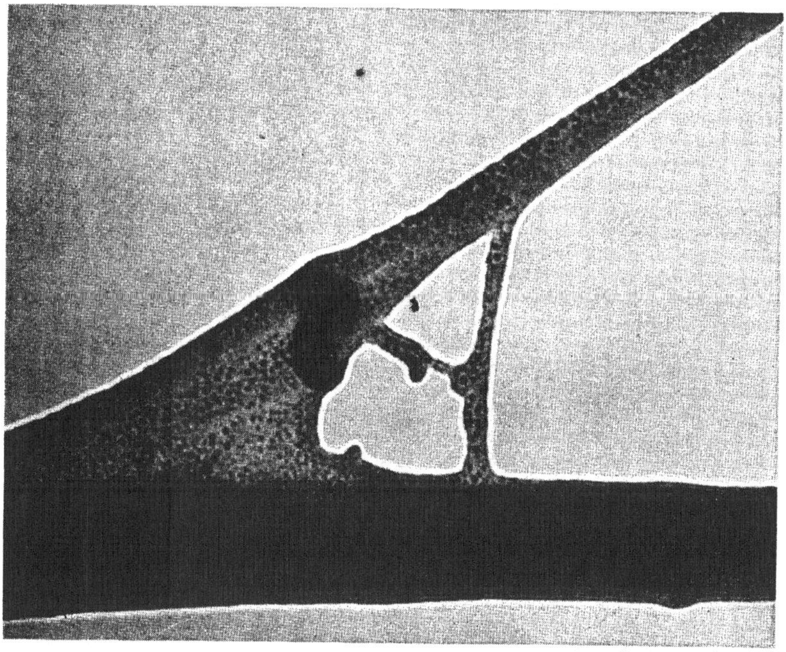

*Radio Corporation of America*

The electron microscope reveals the internal structure of one type of synthetic rubber. The magnification is 225,000 diameters. Although not definitely established the dark spots are not too large to be single molecules.

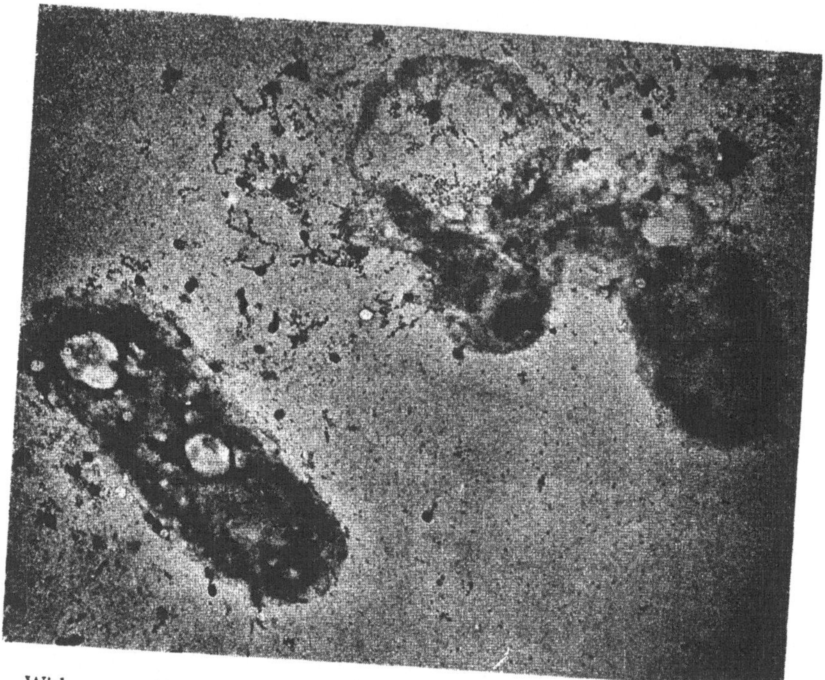

Radio Corporation of America

With a magnification of 28,000 diameters the electron microscope shows the bacillus coli (the elongated objects at upper right and lower left) being destroyed by gamma phages (the small dark spots).

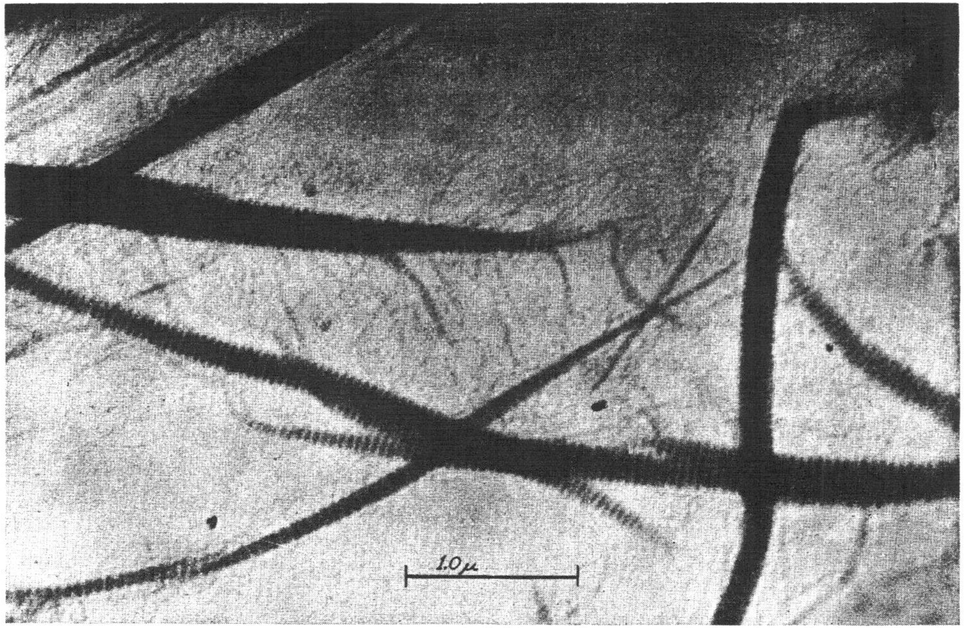


*Dr. F. O. Schmitt and associates at the Massachusetts Institute of Technology*

The chief constituent of tissue and bone is a gelatinous material, called "collagen," which has a long molecule shaped like a corkscrew without the handle. These molecules combine in bundles to form fibers in which the helices lie abreast of each other and likewise the shafts of the corkscrews. The electron microscope with a magnification of 37,500 diameters shows a group of collagen fibers, the dark striations constituting a bundle of the helices and the light striations, the shafts. The scale, marked 1.0$\mu$ or micron, represents a distance of one ten-thousandth of a centimeter.

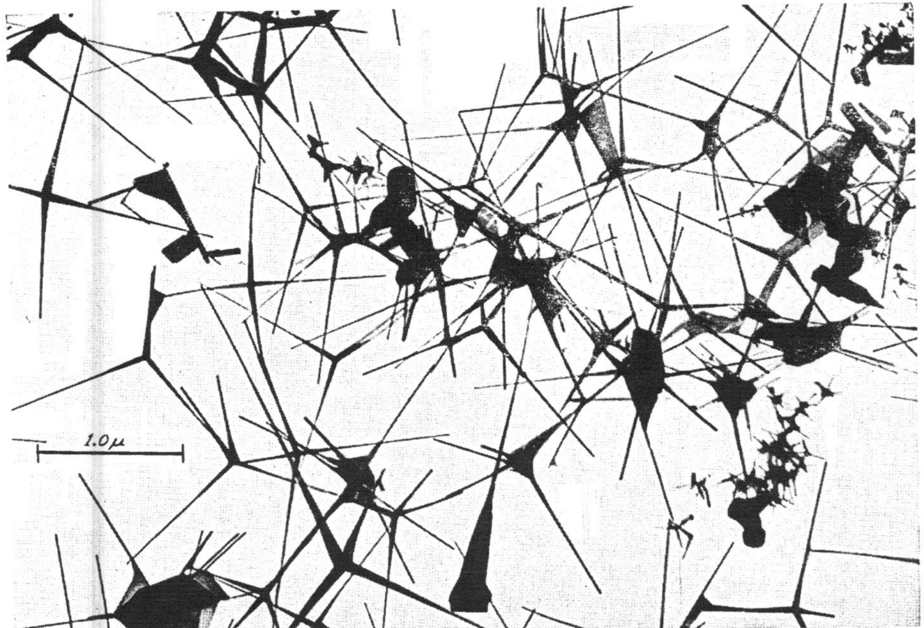


*Dr. F. O. Schmitt and associates at the Massachusetts Institute of Technology*

This is smoke; in particular zinc oxide produced by burning zinc. The electron microscope with a magnification of 34,400 diameters shows the structure of the zinc oxide crystals.

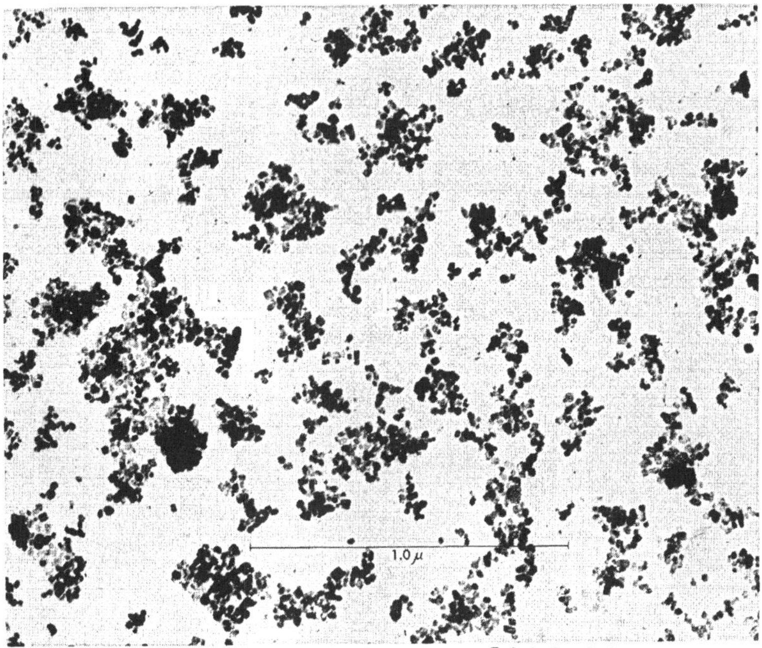


*E. I. du Pont de Nemours and Company*

Colloidal titanium dioxide is a common ingredient of white paint. The photograph taken with an electron microscope with a magnification of 100,000 diameters shows particles which are so small that they must contain only a few thousand molecules. As in all electron-microscope photographs the white particles appear to be black or gray because the electrons are wholly or partially absorbed by the particles and the picture shows only their contours.

*Radiomarine Corporation of America*

A loop antenna, mounted on the pilot house of a ship, which may be turned in any direction from within and indicate the direction of a radio beacon.

*Dewey and Almy Chemical Company*

A radiosonde on its way to the stratosphere. The basket, which is passing the main dome at M.I.T., carries a miniature broadcasting station which sends back periodic records of atmospheric conditions.

*General Electric Company*

A 1,000,000-volt X-ray machine which makes photographic records through thick castings in a few minutes. If any flaws exist, they will be detected.

*General Electric Company*

The 1,400,000-volt X-ray machine in the high-voltage laboratory of the National Bureau of Standards in Washington, D.C. The high voltage is obtained by a series of transformers and electronic rectifiers.

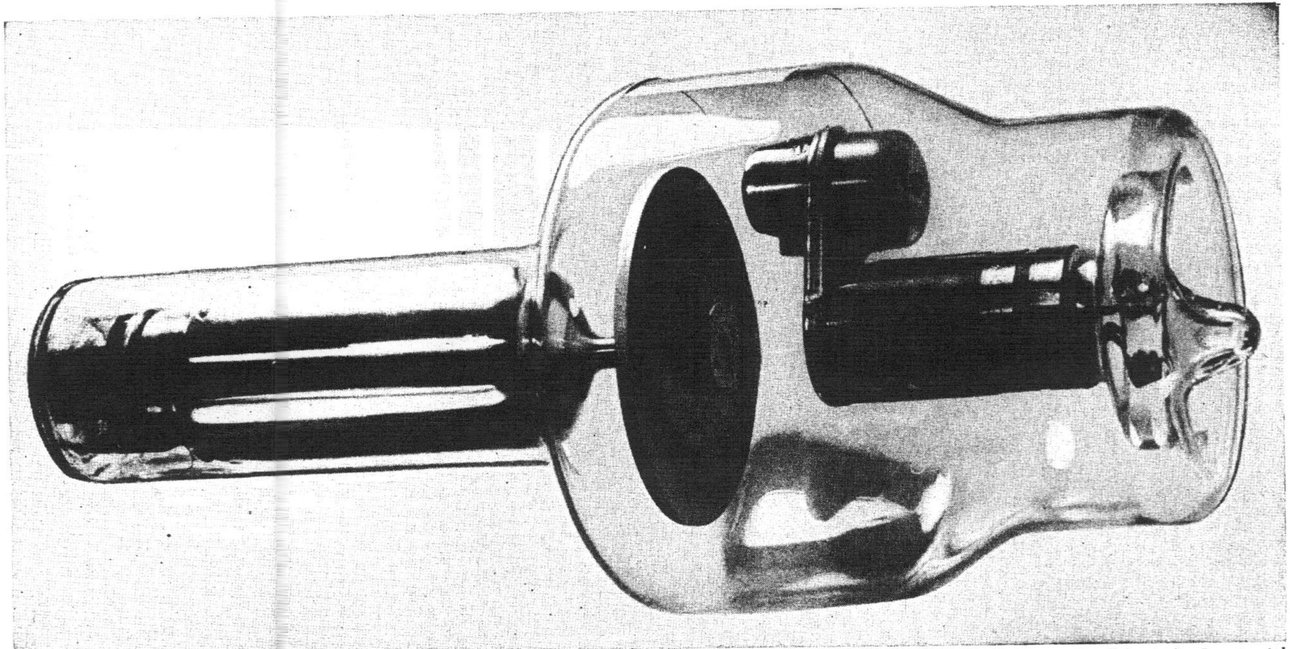

*Machlett Laboratories, Incorporated*

A modern "rotating-targ t" type of diagnostic X-ray tube. A motor with its armature located within the vacuum tube rotates the target at high speed so hat the heat produced by the electron bombardment is distributed over a large surface.

rated at least by the minimum distance required for their resolution. Under the proper conditions two dots separated by only $5 \times 10^{-5}$ inch may be seen through a reading glass as separate dots.

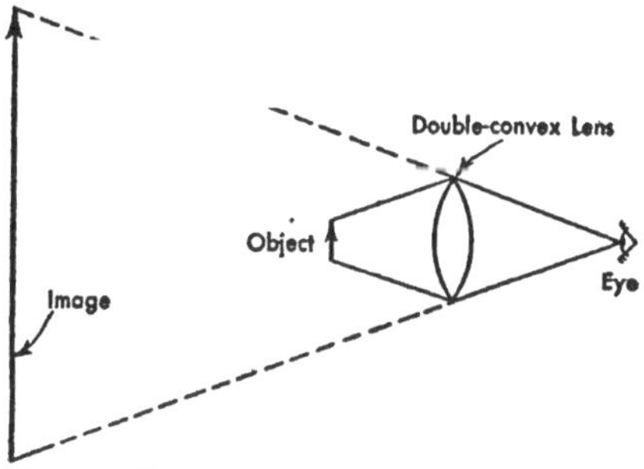


**Fig. 60.** The image of an object is enlarged by a reading glass

With a simple form of compound microscope, as shown in Fig. 61, the object is enlarged by the double-convex objective lens to the first image which is seen by the eye through the ocular lens and is magnified to the second image. The object may be seen by re-

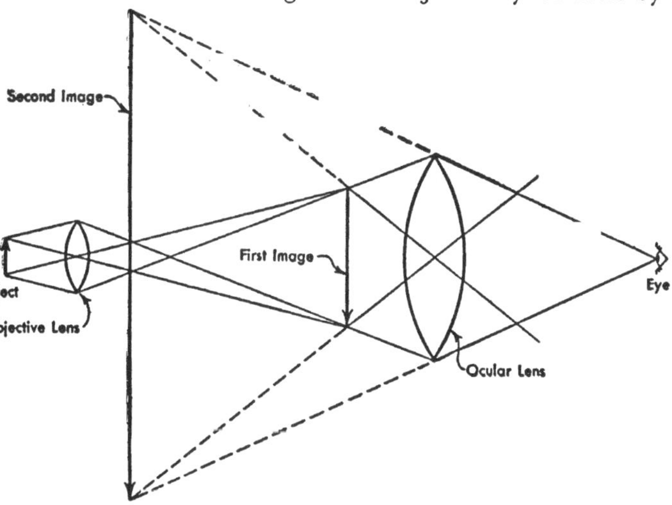


**Fig. 61.** The image of an object is given greater magnification by a compound microscope

flected or transmitted light. The order of visual magnification in the best compound microscope, which may contain as many as ten lenses, is about 500 diameters. With such an instrument dots in the fine structure may be seen as individual dots if they are separated by only $10^{-5}$ inch. Since the shortest visible wave length is about $2 \times 10^{-5}$ inch and the limitations of resolution permit separation only for about half that distance, or $10^{-5}$ inch, it will be seen that visual magnification in excess of 500 diameters will only result in a blurred image. If the image is photographed with an ultraviolet radiation with a wave length of about one quarter that of the shortest visible radiation, the magnification with satisfactory resolution may be extended to 2000 diameters. This brings the detection of the least distance between two dots down to about $2 \times 10^{-6}$ inch. No system of optical lenses is known which will give greater resolution.

It has been predicted theoretically and confirmed experimentally that streams of electrons act like waves of light and, when accelerated by a voltage as high as 100 kilovolts, they may possess wave lengths as small as $2 \times 10^{-11}$ inch or about $10^{-5}$ that of the shortest useful ultraviolet radiation. Optical lenses do not transmit a stream of electrons but either electrostatic or magnetic lenses may be substituted and perform the same function. The electrostatic lenses consist of circular metal plates or disks of various forms with central apertures of different diameters. The magnetic lenses consist of compact toroidal coils of insulated copper wire partially surrounded by an iron sheath. The electron streams are bent by the electrostatic field produced by a high voltage between each pair of plates in the electrostatic type, and by the magnetic field produced by an electric current established in each coil of wire in the magnetic type.

The general features of the electrostatic and the magnetic electron microscope are shown in Figs. 62 and 63. All parts of each microscope are located in a highly evacuated chamber, but in the magnetic type the lenses may be placed outside the chamber for easier adjustment. The object to be magnified is placed on a collodion or a nitrocellulose film about $10^{-6}$ inch thick which is transparent to the electron stream. The magnified image may be viewed on a fluorescent screen or photographed on a plate which replaces the screen. With the best electrostatic or magnetic lenses available at the moment the resulting magnification is about $25 \times 10^3$ diameters. The fluorescent image or the photograph may be enlarged photographically to $10^5$ diameters. With better lenses and higher accelerating voltages the magnification may be increased to $2 \times 10^6$ and perhaps $2 \times 10^8$ diameters.

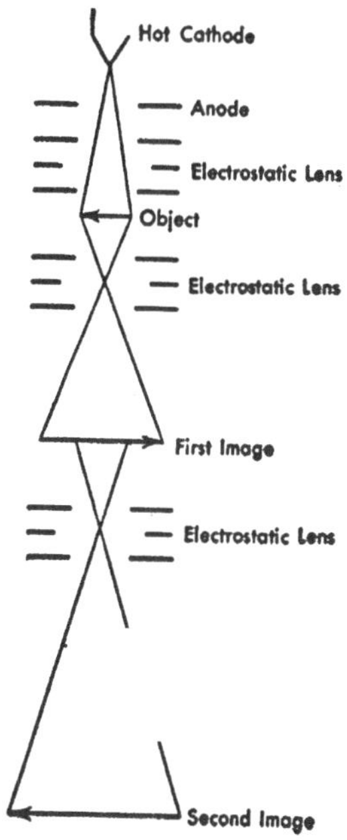


**Fig. 62.** An electron microscope with electrostatic lenses

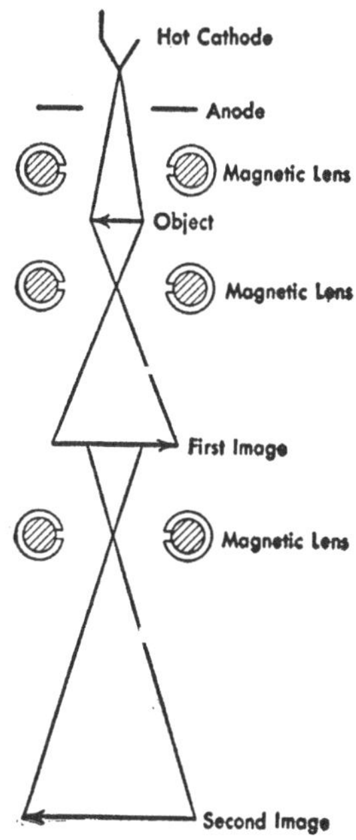


**Fig. 63.** An electron microscope with magnetic lenses

## The Radio Compass

The magnetic or the gyroscopic compass together with the sextant have had extensive application for long-range navigation of ships but are not effective in giving short-range bearings with respect to a fixed object, and the sextant is furthermore incompetent under conditions of low visibility. If two radio beacons, *A* and *B*, send out distinctive, nondirectional signals periodically, the pilot of a ship equipped with a radio compass and any type of terrestrial compass can take alternate bearings on each beacon, as shown in Fig. 64, and plot the successive positions of the ship with respect to the harbor entrance under any condition of weather. If a terrestrial compass is not available or is out of order, the ship's position may still be "fixed" by taking bearings on three radio beacons.

The simplest type of radio compass is the loop antenna, shown in Fig. 65. It consists of a coil, about 15 or 20 inches in diameter, of several turns of insulated wire mounted above the pilot house on

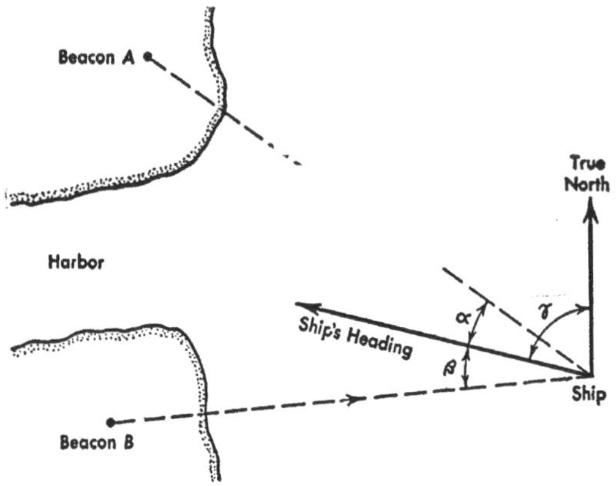


**Fig. 64.** A ship's position may be plotted with respect to two radio beacons

a shaft so that the coil may be turned in any direction from within. The terminals of the coil are connected through a tuning circuit and a vacuum-tube amplifier to a pair of headphones. This type of antenna receives a maximum signal when the plane of the coil points toward the radio beacon and receives a zero signal when the plane of the coil is perpendicular to the intercepted radio wave.

With the plane of the coil pointing toward the radio beacon, the radio wave at a certain instant induces a voltage upward (let us say) at *A* and induces an equal voltage upward but slightly later at *B*. In consequence of the difference in timing a resultant voltage is induced in the coil. If the coil is turned through 90 degrees, the radio wave induces equal voltages upward at *A* and *B* at the same instant and no resultant voltage is induced in the coil. The first difficulty that arises in the application of this device is the "180-degree ambiguity." It will be noted in Fig. 65 that the same signal will be received by the headphones if the radio wave comes from the left or from the right. In other words, on board ship one could not tell whether the radio wave from a single radio beacon is coming from the general direction of the bow or from the stern of the ship.

When employing the maximum-response method one solution of

this problem involves the addition of a vertical antenna in the shaft of the loop antenna. The output of the vertical antenna may be combined or "coupled" with the output of the loop antenna so that the combined signal will be a maximum (let us say) when the radio wave in Fig. 65 comes from the left. When the coil is turned through 180 degrees the signal will be weaker. In this manner it may be determined definitely whether the ship is headed toward the radio beacon or away from it.

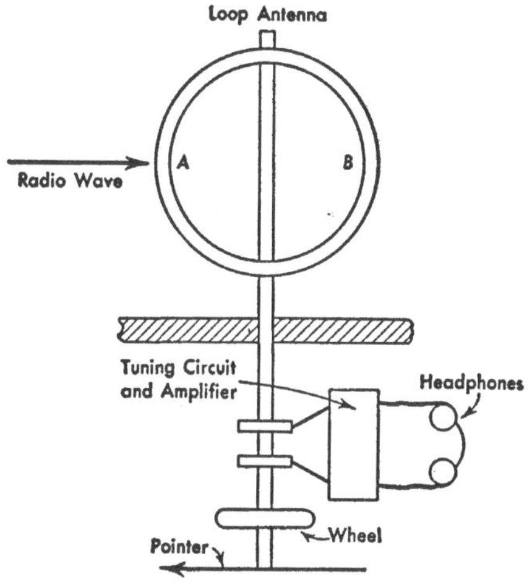


**Fig. 65.** Construction of a radio compass

Under conditions of negligible local noise or static the direction of the radio beacon may be determined more accurately with the minimum than with maximum response; that is, with the plane of the coil perpendicular to the radio wave. This is called the "null method" and other means must be provided for eliminating the 180-degree ambiguity. Let us assume that the radio beacon is located on an island or a lightship which is invisible from the ship. It will be seen from Fig. 66 that when a ship approaches and passes the radio beacon on the left, the radio compass, to maintain the bearing, must be turned steadily in a clockwise direction. If the ship approaches and passes the radio beacon on the right, the radio compass must be turned steadily in a counterclockwise direction. In the case of a single radio beacon on the shore the 180-degree ambiguity may be eliminated in the same manner by changing the ship's course abruptly to the left or right and observing the required rotation of the radio compass to maintain the minimum response.

By either method a ship at sea, receiving the international signal of distress (S O S, ··· ——— ···), will be able to obtain the true bearing and head directly for the ship requesting help. In the case of a lighthouse or lightship which sounds a fog horn at the end of

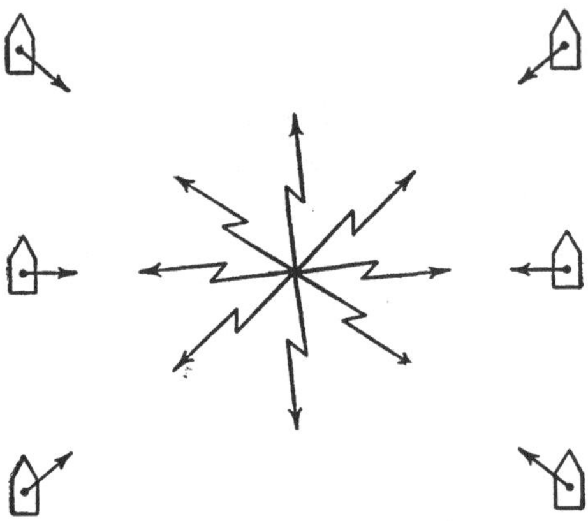

**Fig. 66.** Method of eliminating the 180-degree ambiguity

each radio signal, the distance from the ship to the radio beacon may be determined closely in miles by dividing the time in seconds between the reception of the two signals by 5. On account of the greater velocity of sound in water, if a submarine signal is sent out simultaneously with the termination of the radio signal, the distance of a ship from the radio beacon may be determined in miles by multiplying the time in seconds between the reception of the two signals by 0.78.

While the above description of the operation of the radio compass is discussed with relation to the navigation of ships, the same principles apply to the navigation often called the "avigation," of airplanes. On the airplane the loop antenna is usually located underneath the fuselage and is turned from the cockpit. The effects of static due to higher speed are reduced by enclosing the coil with a nonmagnetic sheath containing a short circumferential airgap or by housing the coil in a streamlined enclosure. With this equipment the pilot of an airplane may either fix his position, aided by a compass, with respect to two radio beacons on land or sea, or turn the loop antenna for maximum or minimum reception and head the airplane toward any desired radio beacon. Radio beacons located at many places throughout the United States send out periodically directional and coded signals which furnish reliable guidance to all pilots within range.

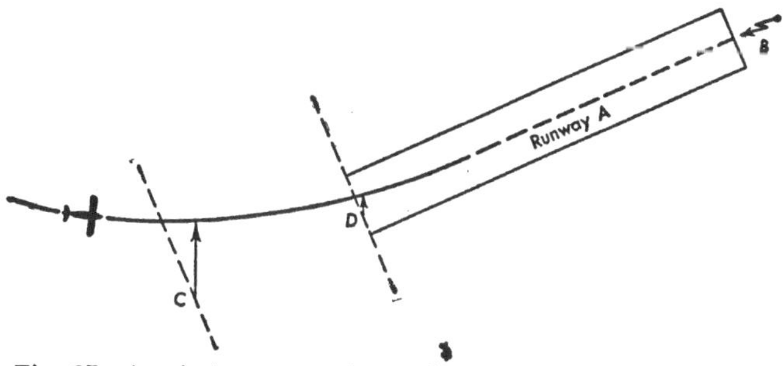


**Fig. 67.** An airplane comes in on the beam and makes a blind landing

Since airplanes move at higher speeds than ships, provision must be made for quicker observations. When the ground cannot be seen the position of the airplane must also be determined in three dimensions so that the pilot will know his height above the ground as well as his horizontal position with respect to the radio beacon. In one form of visual radio compass a pointer on an instrument located on the dashboard swings to the right or the left of the center position when the plane is not headed directly toward the radio beacon. In various types of "blind landing" (Fig. 67) the pilot in an airplane, arriving within a few miles from an airport, may make a landing in the densest fog under the guidance of several instruments. At any height the visual radio compass points the way toward one of the runways, *A*, at the airport, while another pointer informs the pilot whether he is flying above or below a radio beam which rises gradually into the air from a radio beacon, *B*, located at the far end of the runway. Approach markers, *C* and *D*, send up aural or visual signals when the airplane crosses the outer boundary of the airport and again when the airplane crosses the approach end of the runway. Such systems have been perfected which will automatically take over the control of an airplane as it approaches an airport, fly it in "on the beam," and land it without any assistance from the pilot.

Another important aid to aviation is the radio height indicator or absolute altimeter. Former height indicators, which were operated by barometric pressure, gave the altitude above sea level but not above the various parts of a mountain range which might rise in places to a higher elevation than the airplane itself. In the radio altimeter a radio wave directed downward from the airplane is reflected from the earth back to the airplane. The height of the airplane above the ground may be determined by measuring the time required for the double transit of the radio wave at a known velocity. This application of reflected radio waves to focused high-frequency beams will also enable a pilot to explore the region ahead of him as well as below. Pilots of ships at sea by the same principle will be able to determine in the densest fog the distance and direction of another ship or an iceberg in the vicinity and thereby reduce the possibility of collision.

## The Radiosonde

While radio and sound waves have been employed effectively in various ways to explore the depths of the earth's crust for hidden deposits of oil and ore bodies, the physical properties of the atmosphere high above the earth have also been determined accurately from day to day by a balloon-borne device, called a "radiosonde," which steadily transmits the desired information back to a receiving station on the ground. A latex-rubber balloon, weighing about 1.5 pounds and inflated to a diameter of 5 feet, will lift a basket containing meteorological instruments and a miniature broadcasting station, totaling about a pound, to an altitude of 18 miles, and to a lower altitude with a heavier load.

The balloon ascends at a constant velocity of about 1000 feet per minute, expands to a maximum diameter of about 20 feet, and then bursts. The basket is lowered safely to the ground by a parachute. When the basket contains instruments of considerable value two or more balloons are sometimes connected one above the other so that when one bursts the other balloons lower the basket slowly to the ground. Since a balloon is more conspicuous than a parachute this method offers greater probability of recovery of the apparatus.

The power supply for the broadcasting station is obtained from light-weight dry cells or storage batteries with very thin plates. Part of the connection between the balloon and the basket is em-

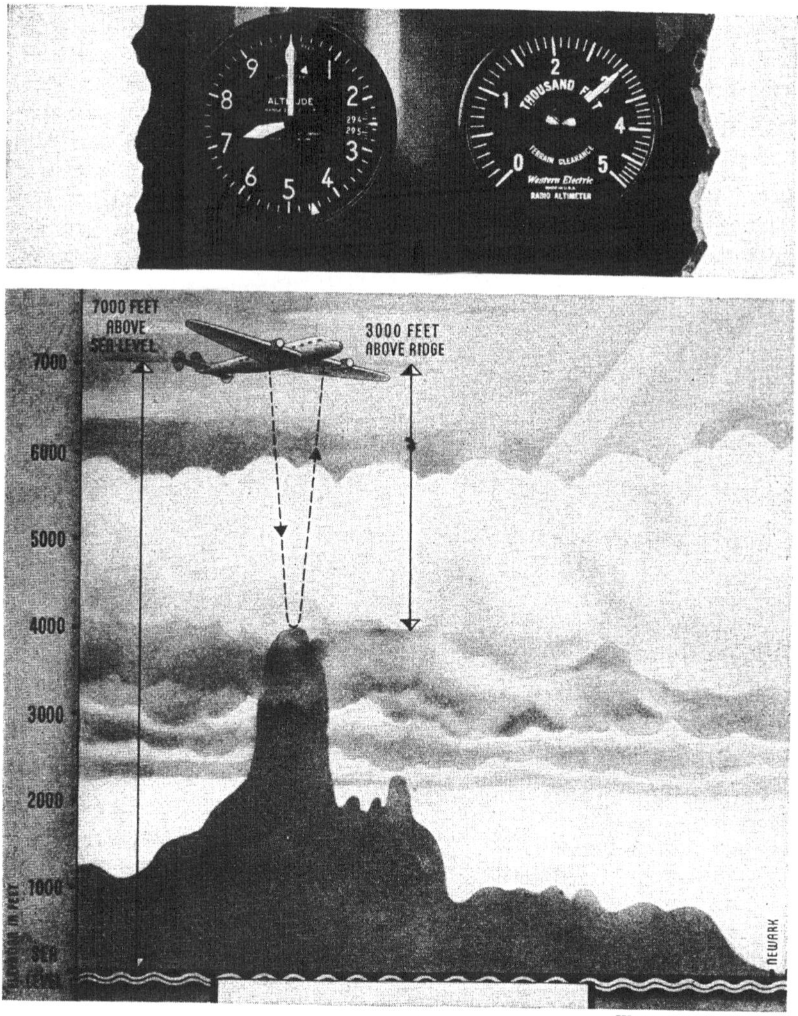


*Western Electric Company*

The "Terrain Clearance" meter (upper right) indicates the height of the airplane above the nearest ground while the barometric altimeter (upper left) indicates the height above sea level. The clearance meter translates to feet one half of the time required for a radio wave of high frequency to reach the ground and return to the airplane. When the clearance is not too great the clearance meter will also indicate, by the vibration of its pointer, the presence of trees or rough ground.

*Western Electric Company*

A close-up of one of the sending and receiving antennas located underneath the wings of an airplane. The height above the nearest ground is determined by the difference in the time required for the radio wave to travel (1) directly from one antenna to the other and (2) from one antenna to the ground and back to the other antenna.

ployed as the broadcasting antenna. For high flights the instruments, batteries, and vacuum tubes are often surrounded by a double layer of cellophane which transmits the intense sunlight to a sufficient degree to keep the apparatus at a temperature of + 50 degrees Centigrade although the surrounding atmosphere may have a temperature as low as − 70 degrees Centigrade. For measurements made at night the apparatus must be enclosed by efficient heat insulation or kept at the proper temperature by the heat generated by water dripping slowly on calcium oxide.

The steady transmission of successive instrument readings back to the ground by radio may be accomplished by various methods. The contactor dial or Olland method is perhaps the simplest and will be explained in terms of the measurement and recording of pressure, temperature, and relative humidity. Suppose the pointer, *OB*, of a pressure gauge within the basket will move in any flight over a maximum scale distance of *A–C* in Fig. 68, the pointer *OD* of a thermometer will move a maximum distance of *C–E* in a second sector of the same dial, and the pointer *OF* of a humidometer will move a maximum distance of *E–A* in the third sector. A contactor arm, *OG*, driven by a small clock or an electric motor, rotates at a speed of two revolutions per minute and makes contact every 30 seconds successively at *A*, *B*, *C*, *D*, *E*, and *F*.

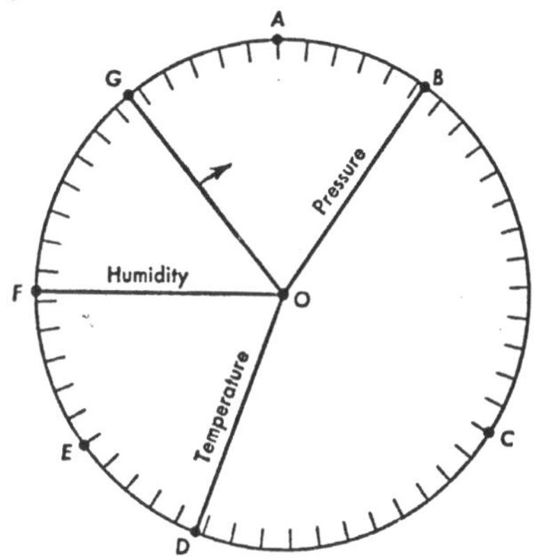


**Fig. 68.** A contactor arm sends radio signals steadily back to ground indicating pressure, temperature, and humidity at various elevations

Each contact sends a radio pulse to the receiving station on the ground where it is recorded on a paper tape moving at constant speed. Assuming that a pulse at *A* indicates zero pressure and a pulse at *C* means 15 pounds per square inch, then a pulse on the tape intermediate between the *A* and the *C* pulse would indicate (for a uniform scale) a pressure of 7.5 pounds per square inch. The temperature and humidity can be read similarly from the tape at half-minute intervals. No error is introduced if the driving clock or motor does not maintain constant speed since a change in speed only changes the rate at which signals are received. If the intensity and frequency of occurrence of cosmic rays are to be measured, a device called a "Geiger counter" installed in the basket will turn on the broadcasting station every time a ray strikes the counter. Pulses of different magnitude on the recording tape will indicate the relative intensities of the intercepted rays and the frequency of their occurrence.

## Medical Electronics

High-frequency radiations, called "X-rays," with a frequency of about $10^{19}$ cps or more, will penetrate or pass through objects which are quite opaque to any radiation of lower frequency. X-rays of moderate frequency are used extensively for the photographic inspection of the internal parts of the body as an aid to diagnosis. It reveals ulcers on teeth, impacted teeth, bone fractures, malformations, foreign bodies, the lesions due to tuberculosis, silicosis, and arthritis, and many other abnormalities within the body. After the administration, orally or by injection, of various solutions less transparent to X-rays the photographs also show outlines of the soft tissues and organs of the body, as well as tumors.

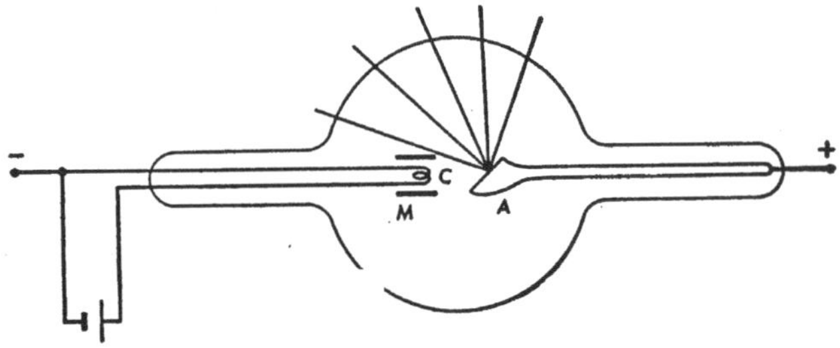


**Fig. 69.** A Coolidge X-ray tube

In the Coolidge X-ray tube shown in Fig. 69 the cathode, *C*, consists of a heated tungsten filament which is surrounded by a molybdenum focusing tube, *M*. The anode, *A*, sometimes called the "anticathode," is made of solid tungsten. The tube is operated at all times under the highest possible vacuum. When a direct voltage of about 100,000 volts is impressed between the anode and the cathode, electrons emitted by the cathode are accelerated to such a high velocity that upon impact with the slanted surface of the anode, a radiation of extremely high frequency is produced. The X-ray emanation from tubes operated at voltages as high as 1,000,000 volts is employed with varying degrees of success for the reduction of malignant growths on the surface or within the body. The same high-intensity X-rays are employed incidentally in the examination of large castings and forgings for internal flaws so that subsequent failure and possible loss of life may be prevented.

In certain types of therapy or healing in which the internal heating of some part of the body or the production of fever in the entire body is indicated, the parts to be treated are surrounded with a few coils of insulated wire conducting a current with a frequency of the order of 1000 kilocycles per second. This high-frequency current is obtained from an oscillator similar to that employed in radio circuits. It induces other secondary currents in the body which will produce under accurate control the proper degree of internal heating. This treatment is called "diathermy" and has had many successful applications.

For many years auscultation, or listening to sounds within the body by means of a stethoscope, has supplied important information to the experienced physician. Major disorders of the heart may be quickly detected by the associated and characteristic sounds but incipient minor defects may pass unnoticed. It has been known for some time that a voltage is produced between two parts of any muscle during contraction. This is particularly true of the heart which also causes similar voltages to pulsate between the extremities of the body. Although such voltages are strong enough to operate sensitive galvanometers directly, the addition of one or more vacuum-tube amplifiers gives more detailed electrocardiagrams, or curves of the motion of the heart.

Contacts are usually made on the patient between the wrists or between one wrist and the opposite ankle. The electrodes consist of small pads moistened with a conducting liquid or glass jars containing a conducting liquid in which the hands and feet are immersed. The current from these electrodes is sent through the moving coil of a galvanometer similar to that employed (page 51) in the production of the sound strip on the Photophone talking-motion picture. A beam of light reflected from a small mirror mounted on the moving coil traces out the heart action on a moving strip of photographic film or photosensitive paper. An electrocardiagram of a typical human heart is shown in Fig. 70. The time between each heart beat at *A* and *B* is 0.8 second. From the detailed structure of each electrocardiagram the experienced examiner will be able to detect any departure from normal and suggest appropriate treatment. Electrocardiagrams also furnish

**Fig. 70.** An electrocardiagram of a typical human heart

desirable records for comparison and study in medical schools. Similar curves produced by the amplified current from a microphone strapped over the heart furnish additional information for cardiac studies and give a permanent record which may be filed for future reference.

Applications of the electrocardiograph to various parts of the body show voltage waves originating from other causes. Of particular interest are the electroencephalograms obtained when the electrodes (moistened pads) are placed on different parts of the skull. These so-called "brain waves" indicate by deviation from normal waves the existence of a psychiatric abnormality and sometimes the presence of a tumor within the brain. Some progress is being made with records made between two electrodes placed anywhere on the body as to its indication of the veracity of statements made by the subject. In this connection the instrument supplying the curve of voltage variations is called a "lie detector."

In electrosurgery the patient is placed on a large metal plate and the surgeon operates with various forms of insulated knife, loop, or needle supplied with a high-frequency current which cuts sharply and at the same time sterilizes and cauterizes the adjacent tissue. This method is particularly effective in surgery of the brain where there must be no chance for a hemorrhage.

A start has been made in a form of electrical diagnosis in which an alternating voltage of moderate frequency (10 kilocycles per second) is impressed between two electrodes placed upon various parts of the body and observation is made of the consequent alternating current which flows through the body between the two electrodes. So far this research has shown that the characteristic response of the patient is not in the magnitude of the current but in the time by which the current lags the impressed voltage. Someday a doctor may apply two moistened sponges to different parts of a patient's body, look at an indicator, and say, "You have diabetes."

## But That Is Not All

Musical instruments are being developed which will reproduce not only all the tonal qualities of an organ or an orchestra but also new tones which have never been heard before. Many of the tones produced by the different pipes of an organ or the instruments of an orchestra have been discovered quite by accident. The tone of a violin, for example, is produced by drawing a bow consisting of the stretched hair of the tail of a white horse over the twisted gut of a sheep of any color stretched between two bridges mounted on a sounding board of critical shape and material. Since the various frequencies of several vacuum-tube circuits may be adjusted accurately to any values and mingled in any relative magnitude and time, the resultant variety of tone is unlimited. The composer of music for future electronic organs or orchestras will combine melody, counterpoint, harmony, and rhythm with a boundless choice of constituent tonal qualities which may be expected to establish a new era in music.

When the voice current in a long-distance telephone line falls below the audible level it is amplified by a group of vacuum tubes, called a "telephone repeater," and sent on to other repeaters. In this manner telephone conversation is made possible between San Diego, California, and Bangor, Maine, a telephone-line distance of 4000 miles requiring 200 repeaters. This line is further extended by radio to London or other foreign cities but a man speaking from San Diego must get down to his office before 9 A.M. because the office in London will be just closing at 5 P.M. In the same manner the voice of a speaker in a large hall or out-of-doors may be amplified and delivered by a loud-speaker to an audience of many thousands. Translators sitting before a speaker in an international meeting may speak into microphones which send amplified currents to earphones installed at each seat in the hall so that each person may turn a dial and listen to the speech in the language with which he is most familiar.

In power plants the smoke density in a chimney may be controlled by a phototube with consequent saving of fuel, and the turbidity of a liquid in a chemical plant may likewise be maintained at the proper value. Manufactured articles without a proper wrapping or label, or defective in shape or color, may be discarded from a belt conveyor. Vegetables, fruits, and nuts may be graded and sorted as to color, shape, or size. Bottles and cans may be filled to the desired volume or weight and capped or sealed. The continuous output of a paper machine may be maintained accurately at a definite thickness and the moisture content of cloth in any textile operation may be kept constant.

The uniform processing of many manufactured materials is dependent upon the maintenance of a constant temperature during the period of formation. This may be accomplished by a succession of workers who must watch a thermometer constantly and regulate

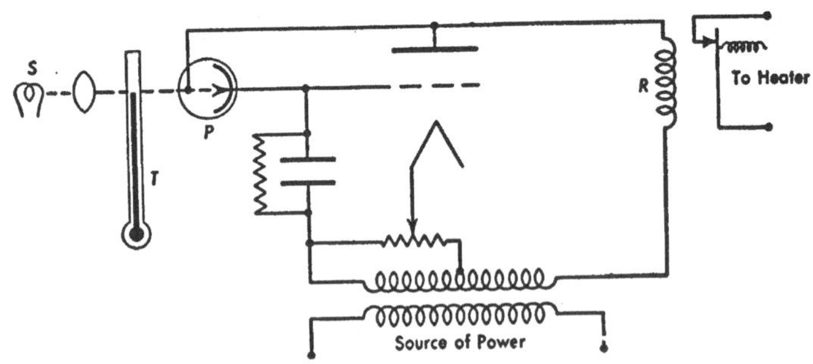


**Fig. 71.** A phototube maintains constant temperature

the source of heat so that there will be no variation in the temperature. Electronic devices will maintain the temperature more accurately and at less expense. In Fig. 71, for example, a small source of light, *S*, is focused through a mercury thermometer, *T*, on a phototube, *P*. The anode of the phototube is connected to the plate of a triode, and the cathode to the control grid. An alternating source of electric power supplies an alternating current to the filament and a rectified current to a relay, *R*, which turns on and off a heater.

Let us suppose that the mercury in the thermometer immersed in the material to be processed falls slightly below the desired level or temperature. The light from the source, *S*, then strikes the cathode of the phototube and raises the control grid to the voltage of the plate. The triode then becomes conducting (see page 31) and a rectified current, flowing through the relay, closes a switch and turns on the heater. An increased supply of heat to the material under temperature control raises its temperature until the rising column of mercury in the thermometer intercepts the light from the source, *S*, stops photoemission in the phototube, and causes the control grid to become sufficiently negative in voltage to interrupt the current in the relay. The supply of heat is cut off until the mercury column in the thermometer drops below the critical point and allows the beam of light again to turn on the heat. A dial thermometer with a hole drilled from face to back at the desired temperature operates in the same manner; that is, when the pointer on the thermometer uncovers the hole the heater is turned on, and is turned off when the pointer again covers the hole. Many other circuits containing electronic devices may be employed which will maintain a temperature to within 0.01 degree Fahrenheit from the desired value.

In the induction furnace, high-frequency currents (20 kc or more) supplied by an oscillator may be employed to induce currents in metallic ores and melt out the metals in a vacuum so that they will not be contaminated by the oxygen, nitrogen, and sulphur of the usual combustion furnace. High-frequency voltages may also be utilized to heat and consolidate nonmetallic materials such as plywood and plastics. Induction cookstoves will cook meat, game, fish, and vegetables from the inside out in a cool oven with a negligible loss of weight and in a matter of seconds rather than minutes or hours. Induction heating will permit people to sit or sleep comfortably without covers in a room with the air temperature below zero.

Phototubes will stop machinery when the product is imperfect. The machinery will also be stopped when an employee endangers any part of his body by reaching or standing too close. Phototubes will detect and trap criminals. They will turn on the lights when the illumination is too low. Show a phototube a sample of cloth or paper and it will tell you, by a curve of the relative intensity of the visible frequencies, exactly what color it is so that you may duplicate that color at any time. Show a phototube an intricate curve inked on a sheet of paper and it will cause a pointer on a machine, called a "servomechanism," to follow that curve over and over again, or cause a piece of material to be cut with the exact shape of the curve. With proper corrections for the drift due to wind and tide a phototube could be shown the desired course of a ship on a map and guide the ship from one port to another.

Runners on a track, horses, automobiles, even bullets traveling 2700 feet per second, may be timed accurately between any two points. The number of drops of lubricating oil supplied to a bearing per minute may be held precisely at any desired value. Objects of any kind manufactured by a machine, people passing through a doorway, and automobiles entering a tunnel may be counted and recorded over any period of time. Mail may be sorted automatically at each post offiee, and merchandise on conveyors may be routed through a factory to its proper destination. Railroad and subway trains may be stopped automatically when they enter a block occupied by another train. An automobile approaching a throughway may be given the green light if the throughway is clear at a safe distance in either direction. The transit of a star across the objective of a telescope may send out a radio signal informing part of the world that it is twelve o'clock.

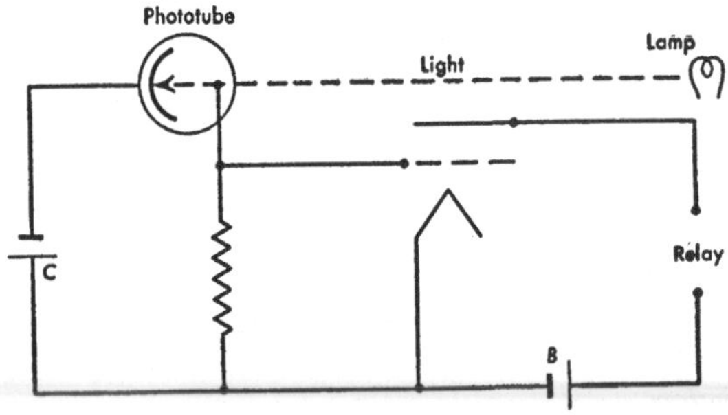


**Fig. 72.** A phototube opens a door

No longer need we push or pull doors; not even say, "Open sesame." Upon approach, a light shining upon the phototube, shown in Fig. 72, is cut off by the intervening body, and the control grid of a triode, which had previously been held by the light at a negative voltage, rises immediately to about the same voltage as the filament. The triode then becomes conducting and the B battery sends current through the relay which admits compressed air to a cylinder. Within the cylinder a piston attached to an arm connected to the door shaft opens the door promptly. Garage doors may be opened by the same method on the approach of an automobile. At eventide the lowing herd need no longer stand patiently at the barnyard gate. It will open at the approach of the first cow. The same principle may be applied to prevent the door of an elevator or a subway train from being closed when a passenger is entering.

A concentrated beam of ultrahigh-frequency electromagnetic waves may be swept steadily across the sky and the sea to determine the location, distance, and motion of airplanes, ships, or surfaced submarines which are otherwise invisible. Upon interception with any object the beam will be reflected back to the source. It can even tell whether the object belongs to a friend or a foe. The development of this indispensable extension of ordinary vision, called "radar," is perhaps the most important contribution to military success in the World War II. It will have equally important applications in peacetime, particularly in the prevention of collisions of airplanes or ships with each other or any other form of obstruction under conditions of low visibility. There will be little need for reduction of speed even in the densest fog.

The electron with its associates in the physical world stands ready to perform any function that the mind of man may dictate. Although its activity, as discussed on the preceding pages, has covered an extensive field, there is no reason to believe that in the centuries to come other applications of fundamental importance will not be discovered which at the moment cannot even be imagined. Electronics is not a realm of science which will bring comfort and pleasure to a few. No person or group of persons may preempt or hoard electrons like gold for their own security and enjoyment. Electrons are the happy and faithful slaves of every man.

# CONVERSION FACTORS

| *Multiply* | *by* | *to obtain* |
|---|---|---|
| Angstroms | $10^{-8}$ | centimeters |
| Atmospheres | 76 | cms of mercury |
| Atmospheres | 14.70 | pounds per sq. inch |
| British thermal units | 778.3 | foot-pounds |
| British thermal units | $2.930 \times 10^{-4}$ | kilowatt-hours |
| Centimeters | $10^{8}$ | angstroms |
| Centimeters | $3.281 \times 10^{-2}$ | feet |
| Centimeters | 0.3937 | inches |
| Centimeters per second | 1.969 | feet per minute |
| Centimeters per second | 0.02237 | miles per hour |
| Cubic centimeters | $6.102 \times 10^{-2}$ | cubic inches |
| Cubic inches | 16.39 | cubic centimeters |
| Degrees Centigrade + 17.8 | 1.8 | degrees Fahrenheit |
| Degrees Fahrenheit − 32 | $\frac{5}{9}$ | degrees Centigrade |
| Dynes | $1.020 \times 10^{-3}$ | grams |
| Dynes | $35.97 \times 10^{-6}$ | ounces |
| Dynes | $2.248 \times 10^{-6}$ | pounds |
| Electron-volts | $1.60 \times 10^{-12}$ | ergs |
| Ergs | $7.378 \times 10^{-8}$ | foot-pounds |
| Foot-pounds | $1.356 \times 10^{7}$ | ergs |
| Foot-pounds | $3.766 \times 10^{-7}$ | kilowatt-hours |
| Grams | 980.7 | dynes |
| Grams | $3.527 \times 10^{-2}$ | ounces |
| Grams | $2.205 \times 10^{-3}$ | pounds |
| Horsepower | 745.7 | watts |
| Inches | 2.540 | centimeters |
| Kilometers | 3281 | feet |
| Kilometers | $10^{3}$ | meters |
| Kilometers | 0.6214 | miles |
| Kilowatts | 737.8 | foot-pounds per sec |
| Kilowatts | 1.341 | horsepower |
| Kilowatts | $10^{3}$ | watts |
| Kilowatt-hours | $2.656 \times 10^{6}$ | foot-pounds |
| Meters | 100 | centimeters |
| Meters | 3.281 | feet |
| Meters | 39.37 | inches |
| Meters | $10^{-3}$ | kilometers |
| Miles | 5280 | feet |
| Pounds | $4.448 \times 10^{5}$ | dynes |
| Pounds | 453.6 | grams |
| Square centimeters | 0.1550 | square inches |
| Square inches | 6.452 | square centimeters |
| Watts | 0.7378 | foot-pounds per sec |
| Watts | $1.341 \times 10^{-3}$ | horsepower |
| Watts | $10^{-3}$ | kilowatts |

# INDEX